Lecture Notes in Mathematics

Edited by A. Dold and B. Ec

T0215974

512

Spaces of Analytic Functions

Seminar Held at Kristiansand, Norway,
June 9–14, 1975

Edited by O. B. Bekken, B. K. Øksendal,
and A. Stray

Springer-Verlag
Berlin · Heidelberg · New York 1976

Editors
Otto B. Bekken
Bernt K. Øksendal
Arne Stray
Agder Distriktshøgskole
Box 607
N-4601 Kristiansand S

Library of Congress Cataloging in Publication Data

Seminar in Functional Analysis and Function
Theory, Kristiansand, Norway, 1975.
Spaces of analytic functions.

(Lectures notes in mathematics ; 512)
Bibliography: p.
Includes index.
1. Analytic functions--Congresses. 2. Function
spaces--Congresses. I. Bekken, O. B., 1940-
II. Øksendal, B. K., 1945- III. Stray, Arne.
IV. Title. V. Series: Lecture notes in
mathematics (Berlin) ; 512.
QA3.L28 no 512 [QA331] 510'.8s [515'.9] 76-7529

AMS Subject Classifications (1970): 30A98, 32E25, 46E15

ISBN 3-540-07682-4 Springer-Verlag Berlin · Heidelberg · New York
ISBN 0-387-07682-4 Springer-Verlag New York · Heidelberg · Berlin

This work is subject to copyright. All rights are reserved, whether the whole or part of the material is concerned, specifically those of translation, reprinting, re-use of illustrations, broadcasting, reproduction by photocopying machine or similar means, and storage in data banks.

Under § 54 of the German Copyright Law where copies are made for other than private use, a fee is payable to the publisher, the amount of the fee to be determined by agreement with the publisher.

© by Springer-Verlag Berlin · Heidelberg 1976

Printed in Germany

Printing and binding: Beltz Offsetdruck, Hemsbach/Bergstr.

PREFACE

This volume contains articles based on talks given at the Seminar on Functional Analysis and Function Theory held at Kristiansand, Norway, June 9 - 14 1975.

Although the papers herein covers a broad area of mathematical research, most of them are in some way connected with algebras and spaces of analytic functions in one or several complex variables. Hopefully this report will be of interest to mathematicians working within these and related areas.

We wish to thank all the participants and contributors for their cooperation. And we gratefully acknowledge the generous financial support from the Norwegian Research Council (NAVF) and from Agder Distriktshøgskole. We also wish to express our gratitude to Professor Erik Alfsen, University of Oslo for his steady encouragement during the preparation of the seminar. We would also like to give special thanks to our secretary Ingrid Skram for her assistance during the seminar and her efficient typing of the manuscripts.

Otto B. Bekken   Arne Stray   Bernt Øksendal

SEMINAR IN FUNCTIONAL ANALYSIS AND FUNCTION THEORY

JUNE 9-14, 1975 IN KRISTIANSAND S., NORWAY

List of participants.

E. Alfsen	Oslo Universitet, Blindern, Oslo 3, Norway
E. Amar	Université de Paris-Sud, Centre D'Orsay 91405, Mathematique, Bât 425, France
E. Andresen	NDH, Box 309, 8001 Bodø, Norway
R. Basener	Lehigh University, Bethlehem, Pennsylvania 18015, USA
O.B. Bekken	ADH, Box 607, 4601 Kristiansand S., Norway
G. Berg	Uppsala Universitet, Sysslomansgatan 8, 75223 Uppsala, Sweden
A. Bernard	Université de Grenoble, BP. 116, 38402 St. Martin D'Heres, France
E. Briem	University of Iceland, 3 Dunhaga, Reykjavik, Iceland
J. Chaumat	Université de Paris-Sud, France
A.M. Chollet	Université de Paris-Sud, France
H.G. Dales	Leeds University, Leeds LS2 9JT, England
A.M. Davie	Edinburgh University, Edinburgh EH 1 1HZ, Scotland
J. Detraz	Université d'Aix-Marseille, 3. Place Victor Hugo, 13331 Marseille Cedex 3, France
G. Dloussky	Université d'Aix-Marseille, France
A. Dufresnoy	Université de Grenoble, France
A. Dvergsnes	MRDH, Box 308, 6401 Molde, Norway
C. Fernstrøm	Uppsala Universitet, Sweden
T.W. Gamelin	UCLA, Los Angeles, California 90024, USA
B. Gaveau	Université de Lille, B.P. 36, 59650 Villeneuve d'Ascq., France
K. Hag	Trondheim Universitet - NTH, 7034 Trondheim, Norway
P. Hag	Trondheim Universitet - NLH, 7034 Trondheim, Norway
A.E. Haugros	ADH, Box 607, 4601 Kristiansand S., Norway
L.I. Hedberg	Stockholms Universitet, Box 6701, S-113 85, Stockholm, Sweden

N.P. Jewell	Edinburgh University, Scotland
S. Kaijser	Uppsala Universitet, Sweden
B. Korenblum	Tel-Aviv University, Ramat-Aviv, Tel-Aviv, Israel
K.B. Laursen	Københavns Universitet, Universitetsparken 5, 2100 København Ø, Denmark
Å. Lima	Box 35, 1432 Ås-NLH, Norway
P. Lindberg	Uppsala Universitet, Sweden
A.M. Mantero	Université de Paris-Sud, France
M. Naghshineh-Ardjmand	Leeds University, England
A. Pełczyński	Polish Academy of Science, Sniadeckich 8, 00950 Warszawa, Poland
R.M. Range	SUNY Albany, New York 12222, USA
J.P. Rosay	Université d'Aix-Marseille, France
D. Sarason	UC Berkeley, Berkeley, California 94720, USA
H.S. Shapiro	Kungl. Tekniska Høgskolan, Matematiska Institutionen, S 100 44, Stockholm, Sweden
N. Sibony	Université de Paris-Sud, France
S.J. Sidney	University of Connecticut, Storrs, Conn. 06268, USA
H. Skoda	Centre Universitaire de Toulon, Château Saint-Michel, 83130 La Garde, France
A. Stray	ADH, Box 607, 4601 Kristiansand S., Norway
J.L. Taylor	University of Utah, Salt Lake City, Utah 84112, USA
N. Varopoulos	Université de Paris-Sud, France
J. Wermer	Brown University, Providence, R.I. 02912, USA
B. Øksendal	ADH, Box 607, 4601 Kristiansand S., Norway
N. Øvrelid	Oslo Universitet, Norway
K. Øyma	ADH, Box 607, Kristiansand S., Norway

LIST OF LECTURES

HILBERT SPACE METHODS AND INTERPOLATING SETS IN THE SPECTRUM OF AN ALGEBRA OF OPERATORS

E.M. Amar

Introduction.

The aim of this note is to study interpolating sets. We prove we can replace algebra interpolation by Hilbert interpolation and using this we prove a theorem on the union of two sets of interpolation.
In the part II we apply this in the case of uniform algebras and using a suitable potential theory we get the necessity of a condition of Carleson type for a set to be interpolating.
In the part III we specialize again and study the quotients of finite dimensions of the disc algebra.

I - Interpolating sets.

a - Let A be a commutative unitary closed sub-algebra of a C^*-algebra. Let M be its Gelfand spectrum and $s \subset M$.

Definition: s is a interpolating set for A if:

$$\exists\, C > 0 \;;\; \forall\, \tilde{s} \subset s \;;\; \text{card}\, \tilde{s} < +\infty \;;\; \forall\, w \in l^\infty(\tilde{s}) \;;$$

$$\exists\, f \in A \text{ s.t. : } \quad \text{i)} \quad \hat{f}(m) = w(m)$$

$$\text{ii)} \quad \|f\|_A \leq C\|w\|_\infty$$

where $\hat{f}(m)$ is the Gelfand transform of f at m.
We see at once that this definition is equivalent to:
the quotient $A/I_{\tilde{s}}$ is isomorphic to $l^\infty(\tilde{s})$, the norm of the isomorphism being less than C. (Here $I_{\tilde{s}}$ is the kernel of $\tilde{s}$ i.e.:

$$I_{\tilde{s}} = \{f \in A \;;\; \hat{f}(m) = 0 \;;\; \forall\, m \in \tilde{s}\}).$$

b - Representations of A.

For studying these quotients we will use representations.
Let r be a cyclic and contracting representation of A acting on the Hilbert space H_r i.e.:

i) $\|r(f)\| \leq \|f\|_A$

ii) $\exists\, h \in H_r$ s.t. : $\{r(f)h \,;\, f \in A\}$ is dense in H_r.

Let m in M and I_m the kernel of m ; because I_m is of codimension one in A the space

$\{r(I_m)h\}^\perp$ is at most of dimension one.

So we have two cases:

i) $\dim\{r(I_m)h\}^\perp = 1$; we pick the unique vector e_m in this space such that: $\|e_m\| = 1$; $(h,e_m) > 0$, where $(\ ,\ )$ is the scalar product in H_r.

ii) $\dim\{r(I_m)h\}^\perp = 0$; by convention we put $e_m = 0$.

Now let $s \subset M$; card $s < +\infty$; and $\pi(f) = r(f)^*$ the involution being taken in $L(H_r)$.
We have:

<u>Lemma</u>. $\forall\, f \in A$; $\forall\, m \in M$: (o) $(r(f)h,e_m) = \hat{f}(m).(h,e_m)$.

(oo) $\pi(f).e_m = \overline{\hat{f}(m)}.e_m$.

<u>Proof</u>. Simple verification.

If we put: E_s = (closed linear span of $\{e_m\}$, $m \in s$)

P_s = the orthogonal projection on E_s, and

$r_s = P_s r$ restricted to E_s

$\pi_s = \pi$ restricted to E_s

we have:

<u>Lemma</u>. r_s is a representation of A/I_s on E_s.
π_s is a <u>diagonal</u> anti-representation of A/I_s on E_s with $\{e_m \,;\, m \in s\}$ as eigenvectors.

Now we have to show that these representations are enough to recapture the norm in A/I_s.
Using the Gelfand-Naimark-Segal construction and a polar decomposition of linear forms on A we can prove, in a very similar manner as B. Cole (1):

Theorem. $\|f\|_{A/I_s} = \sup_{r\in\mathcal{C}} \|r_s(f)\|_{\mathcal{L}(E_s)} = \sup_{\pi\in\mathcal{S}} \|\pi_s(f)\|_{\mathcal{L}(E_s)}$,

where $\mathcal{C}$ is the set of all cyclic and contracting representations of A, and $\mathcal{S}$ the set of all adjoints of such representations.

c - Hilbert interpolating set.

Let H be a Hilbert space and $(e_i \; ; \; i \in L)$ a family of unit vectors in H.

Definition. $(e_i \; ; \; i \in L)$ is an Hilbert interpolating set if

$T : H \to l^\infty(L)$

$Th = \{(h,e_i) \; ; \; i \in L\}$

T is continuous and surjective on $\underline{l^2(L)}$.

This means

i) $\sum_{i\in L} |(h,e_i)|^2 \leq C\|h\|^2$ for all h in H.

ii) $\forall\, w \in l^2(L) \; ; \; \exists\, h \in H$ s.t.: $(h,e_i) = w(i) \;\; \forall\, i \in L$.

W.L.O.G. we can suppose that H is generated by $(e_i)\; i \in L$.

Let $A_L = \{T \in L(H) \; ; \; T.e_i = \hat{T}(i).e_i, \text{ for } i \text{ in } L\}$.
Then A_L is a closed operator algebra of the type of $\pi_s(A)$.
Clearly L may be seen as a part of M, the spectrum of A_L, and we have:

Theorem. $L \subset M$ is an interpolating set for A_L if and only if

$(e_i \; ; \; i \in L)$ is an Hilbert interpolating set.

Using this theorem it remains to characterize the Hilbert interpolation sets. For doing this we need a definition:
if $(e_i \; ; \; i \in L)$ is a family of unit vectors in a Hilbert space H, we say the family $(g_i \; ; \; i \in L)$ forms a biorthonormal system with $(e_i \; ; \; i \in L)$ if

$$\left.\begin{aligned}(g_k,e_i) &= 0 \quad \text{if } k \neq i\\ (g_k,e_k) &= 1\end{aligned}\right\} \quad \text{for all } k,i \text{ in } L$$

We say the system is complete if

$$\forall\, h \text{ in } H \quad h = \sum_{i \in L} (h,e_i)g_i = \sum_{i \in L} (h,g_i)e_i$$

the sums converging in norm. (Of course we have supposed that the combinations of the e_i's are dense in H).
Now we have:

Proposition. The three sentences are equivalent

i) the family $(e_i\ ;\ i \text{ in } L)$ is an Hilbert interpolating set

ii) $\exists\, C > 0$ s.t.: - $\forall\, h$ in H ; $\sum_{i \in L} |(h,e_i)|^2 \leq C\|h\|^2$

- $\exists\ (g_i\ ;\ i \in L)$ forming a biorthonormal system with $(e_i\ ;\ i \text{ in } L)$ with the two properties

- $\|g_i\| \leq C$ all i in L.

- if we put: $e_i' = \dfrac{g_i}{\|g_i\|}$ we must have:

- $\forall\, h$ in H ; $\sum_{i \in L} |(h,e_i')|^2 \leq C\|h\|^2$

iii) there exists $(g_i\ ;\ i \text{ in } L)$ forming with $(e_i\ ;\ i \in L)$ a complete biorthonormal system.

Using this proposition and estimates based on the geometry of the unit sphere of an Hilbert space, we can prove:

Theorem. Let s_1 and s_2 be two interpolating sets for A such that the Gleason distance between two points of $s_1 \cup s_2$ is uniformly bounded from below. Then $s_1 \cup s_2$ is an interpolating set for A.

This theorem generalizes to operator algebras a theorem of N.Th. Varopoulos, stated in the case of uniform algebras (2). Actually, in this case Varopoulos treated the union of two compact interpolating sets.

II - Uniform algebra case.

Let A now be a uniform algebra ; M its spectrum and λ a probability measure on M.
Let $H^2(\lambda)$ be the closure of A in $L^2(\lambda)$ and r the multiplication representation of A on $H^2(\lambda)$.
Taking 1, the unit of A, as cyclic vector and using the preceding notations, for each m in M we put

$$\rho(m) = (1, e_m) \geq 0$$

$$M^\lambda = \{m \text{ in } M \text{ ; s.t.: } \rho(m) \neq 0\}.$$

Then we have (part I)

$$\forall f \text{ in } A \text{ ; all } m \text{ in } M^\lambda : (f, e_m) = \rho(m).\hat{f}(m)$$

$$: \pi(f).e_m = \overline{\hat{f}(m)}.e_m.$$

where, as usual, π is the adjoint of r.

a) Poisson kernel and potential structure on M^λ.

Let $P_m = |e_m|^2$ for all m in M^λ.
Then we have: - P_m is in $L^1(\lambda)$ and $\|P_m\|_1 = 1$.
- P_m is positive
- for all f in A ; all m in M^λ : $\hat{f}(m) = \int_M fP_m \, d\lambda$

For all h in $H^2(\lambda)$ we put: $\hat{h}(m) = (h, e_m) \frac{1}{\rho(m)}$ for m in M^λ.
This definition is compatible a.e. (λ) as easily seen and allows us to put

$$P_m(m') = |\hat{e}_m(m')|^2 \quad \text{for } m, m' \text{ in } M^\lambda.$$

Now we are in position to give:

Definition. We call cell of "center" m and aperture $t > 0$, the set

$$C_{m,t} = \{m' \text{ in } M^\lambda \text{ ; s.t. : } \rho(m)^2 P_m(m') > 1/t\}$$

We call ball of "center" m and aperture t

$$B_{m,t} = C_{m,t} \cap \mathrm{supp}(\lambda).$$

We call pseudo-distance to the boundary

$$d(m) = \rho(m)^2.$$

these definitions are the classical ones when we have a Poisson kernel.

Now we return to interpolation.

b) Application to $H^2(\lambda)$ interpolating sets.

Let $s \subset M$ such that: $s \cap M^\lambda = s^\lambda$ is an $H^2(\lambda)$ interpolating set, i.e.: $(e_m\ ;\ m \text{ in } s^\lambda)$ is Hilbert interpolation. For instance if s is interpolating for A then s^λ is $H^2(\lambda)$ interpolating for all λ.

We put: $\mu = \sum_{m \in s^\lambda} d(m).\delta_m$ where δ_m is the Dirac measure at the point m. Then μ is a positive measure and, as an easy consequence of part I, we have:

Theorem. If s^λ is $H^2(\lambda)$ interpolation, then we have

$$\mu[C_{m,t}] \leq C.t.d(m).$$

If the family of kernel is uniformly trunkable by $d(m)$: i.e.: $\int_{B_{m,b}} P_m \geq 1 - a(t)$; where $a(t)$ depends only of t and $\lim a(t) = 0$ when $t \to \infty$. Then $d(m) \leq t.\lambda[B_{m,t}]$ and then we obtain

$$\mu[C_{m,t}] \leq C.t^2.\lambda[B_{m,t}].$$

This means exactly that the interpolation measure μ must be a "Carleson" measure with respect to the given measure λ.(3,4,5).

III - Disc algebra case.

We apply these kind of ideas to the disc algebra. We take as the given measure the Lebesgue measure on the torus T. Let s be a finite set in $D = (z \in C, |z| < 1)$, and I_s the kernel of s.

In this case we have quite a miracle:

Theorem. π_s is an isometric anti-linear representation of A/I_s.

This allows us to characterize the unit suphere of A/I_s:

Theorem. The unit sphere of A/I_s is exactly made by equivalence classes of Blaschke products of at most "card s -1" zeros.

And if s has n points:

Corollary. The best interpolating function is a constant time a Blaschke product of at most n-1 zeros.

Following this line of ideas we can give a new proof of the well known theorem of Carleson: (6).

Theorem. s is an interpolating set for $H^\infty(D)$ if and only if the product of Gleason distance is uniformly bounded from below.

The first theorem stated here says, in particular, that H^2 interpolation is the same as H^∞ interpolation. This is no longer true for the polydisc or the ball algebra, as was proved by Denise, Amar and Sarroste.
In this respect, disc algebra is very peculiar.

REFERENCES.

[1] J. Wermer, B. Cole: Quotient algebras of uniform algebras. Symposium of Uniform Algebras and Rational Approximation. University of Michigan, 1969.

[2] N.Th. Varopoulos: C.R.A.S. Série A, t. 272, p. 950, 1971.

[3] N.Th. Varopoulos: C.R.A.S. Série A, t. 274, p. 1539, 1972.

[4] L. Carleson: An interpolation problem for bounded analytic functions. Amer. J. Math. 80, 1958.

[5] L. Carleson and J. Garnett: Interpolating sequences and separation properties. Preprint.

[6] L. Carleson: Interpolation by bounded analytic functions and the Corona problem. ANN. Math. 76, 1962.

EXTREME ORTHOGONAL BOUNDARY MEASURES FOR A(K) AND DECOMPOSITIONS FOR COMPACT CONVEX SETS

Eggert Briem

Introduction. If $A = A(K)$ denotes the space of all continuous affine functions on a compact convex set K it is proved that the set $A_1^{\perp} \cap M(\partial K)$ of all boundary measures in the unit ball $A_1^{\perp}$ of all measures orthogonal to A intersects the set $\partial_e A_1^{\perp}$ consisting of the extreme points of $A_1^{\perp}$. Further, it is proved that $A_1^{\perp} \cap M(\partial K)$ is the closed convex hull of its set of extreme points $\partial_e A_1^{\perp} \cap M(\partial K)$, in the weak topology defined on $A^{\perp}$ by the linear span of $C(K)$ and the set $\{f.\hat{g} : f,g \in C(K)\}$. This is a sort of a parallel to the result that $A_1^{\perp}$ is the closed convex hull of its set of extreme points in the topology defined on $A^{\perp}$ by $C(K)$. The above stated result may also be viewed as a parallel to the result in ([1], Theorem I.6.14.) which says that for a given point $x \in K$, there are boundary measures among the extreme points $\partial_e M_x^+(K)$ of the set $M_x^+(K)$ of all positive measures on K representing x, and the set $M_x^+(\partial K)$ of all boundary measures in $M_x^+(K)$ is the closed convex hull of its set of extreme points $\partial_e M_x^+(K) \cap M(\partial K)$, in the weak topology defined on $M(K)$ by the linear span of $C(K)$ and the set $\{\hat{f} : f \in C(K)\}$.

In the second part of this note the result stated above is used to study facial decompositions for K mainly focusing on the Bishop decomposition for K introduced and investigated in [4] and [5].

I. Extreme orthogonal boundary measures.

Let K be a compact convex subset of a locally convex Hausdorff space and let ∂K denote the set of extreme points for K. By $A = A(K)$ we denote the Banach space of all continuous real-valued affine functions on K; by M(K) the Banach space of all Radon measures on K and by $M(\partial K)$ the subspace of M(K) consisting of those $\mu \in M(K)$ for which the total variation measure $|\mu|$ is maximal in Choquet's ordering, or equivalently for which the positive part μ^+ and negative part μ^- are both maximal in Choquet's ordering. The elements of $M(\partial K)$ are called boundary measures. Further, $M_1^+(K)$, resp. $M_1^+(\partial K)$ denotes the set of probability measures in M(K), resp. $M(\partial K)$. The set of all measures in M(K) orthogonal to A is denoted by $A^\perp$, its unit ball by $A_1^\perp$ and the set of extreme points for $A_1^\perp$ is denoted by $\partial_e A_1^\perp$. Thus $A_1^\perp \cap M(\partial K)$ denotes the set of all boundary measures of norm at most one which are orthogonal to A. The sets $A_2^\perp$ and $\partial_e A_2^\perp$ are defined analogously.

If $f \in C(\overline{\partial K})$ we define the upper envelope $\hat{f}$ by $\hat{f}(x) = \inf \{a(x) : a \in A \text{ and } a|_{\overline{\partial K}} > f\}$, and the lower envelope $\check{f}$ by $\check{f}(x) = \sup \{a(x) : a \in A \text{ and } a|_{\overline{\partial K}} < f\}$.

Finally, we let P(K) denote the convex cone of all continuous and convex real-valued functions on K.

The set $M_1^+(K)$ is a w*compact convex subset of M(K); we look at the set $M_1^+(K) \times M_1^+(K)$ equipped with the product topology; it is a convex compact set. We let M denote the subset of $M_1^+(K) \times M_1^+(K)$ defined as follows:

$$M = \{(\mu,\lambda) : (\mu,\lambda) \in M_1^+(K) \times M_1^+(K) \text{ and } \mu - \lambda \in A^\perp\}$$

M is clearly a convex compact subset of $M_1^+(K) \times M_1^+(K)$. In M (or more generally in $M_1^+(K) \times M_1^+(K)$) we introduce the ordering of Choquet in each component i.e. $(\mu',\lambda') \geq (\mu,\lambda)$ if

$$\int f d\mu' \geq \int f d\mu \text{ and } \int f d\lambda' \geq \int f d\lambda$$

for all $f \in P(K)$. Then M is an ordered convex compact. Let Z = Z(M) denote the subset of M consisting of those $(\mu,\lambda) \in M$ which are maximal in the ordering defined on M; this is clearly the set of those $(\mu,\lambda) \in M$ for which both μ and λ are maximal in Choquet's ordering. If $\partial_e M$ denotes the set of extreme points for M we let $\partial_e Z$ denote the set $\partial_e Z = Z \cap \partial_e M$.

PROPOSITION 1. $\partial_e Z \neq \emptyset$

PROOF. Let $f,g \in P(K)$ and put

$$\alpha = \sup \{ \int f d\mu + \int g d\lambda : (\mu,\lambda) \in M \}$$

If $(\mu_n,\lambda_n) \in M$ $n=1,2,\ldots$ and if

$$\lim_{n\to\infty} (\int f d\mu_n + \int g d\lambda_n = \alpha$$

then, if (μ,λ) is a cluster point of $\{(\mu_n,\lambda_n)\}$ we see that

$$\int f d\mu + \int g d\lambda = \alpha.$$

Thus, the set

$$F = \{(\mu,\lambda) \in M: \int f d\mu + \int g d\lambda = \alpha \}$$

is nonempty. F is a closed face of M and if $(\mu,\lambda) \in M$ and $(\mu,\lambda) \geq (\mu',\lambda') \in F$ then $(\mu,\lambda) \in F$ so that F is hereditary upwards. But then $F \cap \partial_e Z \neq \emptyset$([1], Prop. 1.6.4).

Now look at the weak topology σ defined on $M(K) \times M(K)$ by functionals of the form

$$(\mu,\lambda) \to \int f d\mu + \int g d\lambda$$

where $f,g \in D(K)$, the linear span of $C_R(K)$ and $\{ \hat{f} | f \in C(K)\}$.

THEOREM 2. Z <u>is the closed convex hull of</u> $\partial_e Z$ <u>in the weak topology</u> σ <u>defined above on</u> $M(K) \times M(K)$.

PROOF. If μ is maximal then

$$\int (f-\hat{f})d\mu = 0$$

for all $f \in C_R(K)$ and this characterizes the maximal measures. Therefore Z is closed in the topology σ. If $(\mu',\lambda') \in Z \setminus \sigma\text{-cl.conv.}\ \partial_e Z$ then we can find f and g in D(K) such that

$$\sup \{ \int f d\mu + \int g d\lambda : (\mu,\lambda) \in \partial_e Z \} = \alpha < \int f d\mu' + \int g d\lambda'$$

Now,

$$f = f_o + \sum_{i=1}^{n} t_i \hat{f}_i , \quad g = g_o + \sum_{j=1}^{m} s_j \hat{g}_j$$

Since all measures involved are boundary measures we see that if

$$f' = f_o + \sum_{i=1}^{n} t_i f_i \quad \text{and} \quad g' = g_o + \sum_{j=1}^{m} s_j g_j \quad \text{then}$$

$$\sup \{ \int \check{f} d\mu + \int \check{g} d\lambda : (\mu,\lambda) \in \partial_e Z \} = \alpha < \int \check{f} d\mu + \int \check{g} d\lambda .$$

Since

$$\int \check{h} d\xi = \sup \{ \int p d\xi : p \in P(K)\ p \leq \check{h} \}$$

there exists $f'', g'' \in P(K)$ such that

$$\sup \{ \int f'' d\mu + \int g'' d\lambda : (\mu,\lambda) \in \partial_e Z \} \leq \alpha < \int f'' d\mu' + \int g'' d\lambda'$$

But the proof of Proposition 1 shows that for some $(\mu,\lambda) \in \partial_e Z$ we have

$$\int f'' d\mu + \int g'' d\lambda \geq \int f'' d\mu' + \int g'' d\lambda'$$

and we have reached a condradiction.

Let us look more closely at $\partial_e Z$.

LEMMA 3. <u>If</u> $(\mu,\lambda) \in \partial_e Z$ <u>then either</u> $\mu = \lambda$ <u>or</u> $||\mu-\lambda|| = 2$.

PROOF. Suppose $\mu \neq \lambda$ and $||\mu-\lambda|| < 2$. We decompose $\mu-\lambda$ into its positive and negative part: $\mu-\lambda = \xi^+ - \xi^\div$. Then $||\mu-\lambda|| = ||\xi^+|| + ||\xi^\div|| = 2||\xi^+|| = 2||\xi^\div||$. (the last two equality signs hold because $\xi^+ - \xi^\div = \mu-\lambda \in A^\perp$). Put $\eta = \lambda - \xi^\div = \mu - \xi^+$. The measure η is positive, this follows from the fact that ξ^+ and $\xi^\div$ live on disjoint subset of K, and $\mu = \xi^+ + \eta$, $\lambda = \xi^\div + \eta$ and hence $1 = ||\xi^+|| + ||\eta|| = ||\xi^\div|| + ||\eta||$. But if we put $\alpha = ||\xi^+||$, $\mu' = \xi^+/_{||\xi^+||}$, $\lambda' = \xi^\div/_{||\xi^\div||}$ and $\mu'' = \lambda'' = \eta/_{||\eta||}$ then (μ', λ') and (μ'', λ'') are in M, $\alpha \in (o,1)$ and $(\mu,\lambda) = \alpha(\mu',\lambda') + (1-\alpha)(\mu'',\lambda'')$, contradicting the fact that $(\mu,\lambda) \in \partial_e Z$. Thus, either $\mu = \lambda$ or $||\mu-\lambda|| = 2$.

REMARK. We note that if K is a simplex then $\mu = \lambda$ for all $(\mu,\lambda) \in Z$, (in particular for all $(\mu,\lambda) \in \partial_e Z$) and that if $\mu = \lambda$ for all $(\mu,\lambda) \in \partial_e Z$ then $\mu = \lambda$ for all $(\mu,\lambda) \in Z$ so that $A^\perp \cap M(\partial K) = \{0\}$ i.e. K is a simplex.

LEMMA 4. <u>If</u> K <u>is not a simplex then</u>

$$\partial_e A_2^\perp \cap M(\partial K) = \{ \mu-\lambda \mid (\mu,\lambda) \in \partial_e Z \text{ \underline{and} } ||\mu-\lambda|| = 2 \}$$

PROOF. Let $\xi \in \partial_e A_2^\perp \cap M(\partial K)$, then $||\xi|| = 2$ and we can write $\xi = \xi^+ - \xi^\div$ where ξ^+ and $\xi^\div$ are maximal measures with $||\xi^+|| = ||\xi^\div|| = 1$ so that $(\xi^+,\xi^\div) \in Z$. But clearly $(\xi^+, \xi^\div) \in \partial_e Z$. Conversely if $(\mu,\lambda) \in \partial_e Z$ and $||\mu-\lambda|| = 2$, then $\xi = \mu-\lambda \in A_2^\perp$ and if $\xi = \alpha\xi_1 + (1-\alpha)\xi_2$ for some $\alpha \in (o,1)$ and $\xi_1, \xi_2 \in A_2^\perp$ then $||\xi|| = ||\xi_1|| = ||\xi_2|| = 2$. But then $\xi^+ = \alpha\xi_1^+ + (1-\alpha)\xi_2^+$ and $\xi^\div = \alpha\xi_1^\div + (1-\alpha)\xi_2^\div$ so that $(\mu,\lambda) = (\xi^+,\xi^\div) = \alpha(\xi_1^+,\xi_1^\div) + (1-\alpha)(\xi_2^+,\xi_2^\div)$. Hence $\xi = \xi_1 = \xi_2$ and we conclude that $\mu-\lambda \in \partial_e A_2^\perp \cap M(\partial K)$.

THEOREM 5. $A_1^{\perp} \cap M(\partial K)$ <u>is the closed convex hull of</u> $\partial_e A_1^{\perp} \cap M(\partial K)$ <u>in the topology</u> τ <u>defined on</u> $M_R(K)$ <u>by</u> $\tilde{D}(K)$, <u>the linear span of</u> $C(K)$ <u>and</u> $\{ g\hat{f} : g,f \in C(K) \}$.

PROOF. Suppose $\xi \in M(K)$ and that (ξ_α) is a net from $A_1^{\perp} \cap M(\partial K)$ converging to ξ in the topology τ. Clearly $\xi \in A_1^{\perp}$. Let us show that $\xi \in M(\partial K)$. To do so it suffices to show that

$$\int (\hat{f}-f)d|\xi| = 0$$

for all $f \in C(K)$. Given $\varepsilon > 0$ pick a continuous function h on K such that $\| |\mu|-h\mu \| < \varepsilon$. Now if f is any continuous function on K, with $\|f\| = 1$ we have:

$$0 \leq \int (\hat{f}-f)d|\mu| \leq \int (\hat{f}-f)hd\mu + \varepsilon = \lim_\alpha \int (\hat{f}-f)hd\xi_\alpha + \varepsilon = \varepsilon$$

The set $A_1^{\perp} \cap M(\partial K)$ is thus closed in the topology τ. Suppose there is a measure $\eta \in A_2^{\perp} \cap M(\partial K)$, $\|\eta\| = 2$, but $\eta \notin \tau$-cl.conv. $(\partial_e A_2^{\perp} \cap M(\partial K))$. Then there is a function $f \in \tilde{D}(K)$ such that

$$\sup \{ \int fd\xi : \xi \in \partial_e A_2^{\perp} \cap M(\partial K) \} = \alpha < \int fd\eta .$$

By Lemma 3 and Lemma 4

$$\sup \{ \int fd\mu + \int -fd\lambda : (\mu,\lambda) \in \partial_e Z \} = \alpha < \int fd\eta^+ + \int -fd\eta^- .$$

Since all measures occuring above are boundary measures this leads to a contradition, as in the proof of the previous theorem. Thus, $A_2^{\perp} \cap M(\partial K) \subseteq \tau$-cl.conv. $(\partial_e A_2^{\perp} \cap M(\partial K))$ or, equivalently $A_1^{\perp} \cap M(\partial K) \subseteq \tau$-cl.conv. $(\partial_e A_1^{\perp} \cap M(\partial K))$.

II. Decompositions for K

We call a family $\{F_\alpha\}$ of pairwise disjoint split faces of K covering ∂K a decomposition for K. The question is whether the family $\{F_\alpha\}$ determines $A(K)$ in some sense. F.ex. when the family

$\{F_\alpha\}$ consists of the sets of constancy for the central functions in A(K) then $A(K)|_{\overline{\partial K}} = \{ f \in C(\overline{\partial K}) : f|_{F_\alpha \cap \overline{\partial K}} \in A(F_\alpha)|_{F_\alpha \cap \overline{\partial K}} \}$ ([4], Theorem 2). If K is compact convex set let us denote by B(K) the set $B(K) = \{ f \in C(\overline{\partial K}) : f$ is annihilated by every boundary measure orthogonal to A(K) $\}$. For a given decomposition $\{F_\alpha\}$ of K one might ask, whether $B(K) = \{ f \in C(\overline{\partial K}) : f|_{\overline{\partial F_\alpha}} \in B(F_\alpha)$ for each $\alpha \}$. This turns out to be true when $\{F_\alpha\}$ is the Bishop decomposition for K introduced by Ellis in [4]. To prove this we need a lemma.

LEMMA 6. <u>Let</u> $\mu \in \partial_e A(K)_1^\perp \cap M(\partial K)$ <u>and let</u> F <u>be the smallest closed split face of</u> K <u>containing the support of</u> μ. <u>Then the center of</u> A(F) <u>is trivial i.e. contains only the constant functions</u>.

PROOF. By definition the center of A(F) consists of those functions f in A(F) such that for each $a \in A(F)$ there is a function $b \in A(F)$ with $a(x)f(x) = b(x)$ for all x in ∂F and hence also for all x in $\overline{\partial F}$. Thus, if f is in the center of A(F) and $0 \leq f \leq 1$ then $f\mu$ and $(1-f)\mu \in A(K)^\perp$. This shows that f must be constant on the support of μ. Since sets of constancy for f are split faces of F and hence split faces of K ([3], Theorem 1) the minimality of F shows, that f is constant on F.

In [4] Ellis proved that there is a family $\{F_\beta\}$ of pairwise disjoint split faces of K covering ∂K such that for each β, $A(F_\beta)$ has trivial center and such that if E is a subset of ∂K for which $A(\overline{co}(E))$ has trivial center the E is contained in some F_β. This family $\{F_\beta\}$ is called the Bishop decomposition for K. Lemma 6 and the maximality properties of the sets F_β show that the support of each $\mu \in \partial_e A(K)^{\perp}_1 \cap M(\partial K)$ is contained in some F_β. From Theorem 5 we thus get:

THEOREM 7. Let $\{F_\beta\}$ be the Bishop decomposition for K then

$$\{ f \in C(\overline{\partial K}) : \int f d\mu = 0 \text{ for all } \mu \in A(K)^{\perp} \cap M(\partial K) \} =$$

$$\{ f \in C(\overline{\partial K}) : \int f d\mu = 0 \text{ for all } \mu \in A(F_\beta)^{\perp} \cap M(\partial F_\beta) \text{ and for all } \beta \}.$$

COROLLARY: If each F_β is a singleton then K is a simplex.

By ([2] Theorem 1), a function f on $\overline{\partial K}$ extends to a function in A(K) if and only if $f = \hat{f} = \check{f}$ and f is annihilated by all $\mu \in A(K)^{\perp} \cap M(\partial K)$. Thus, the Bishop decomposition determines K within the set $\{ f \in C(\overline{\partial K}) : f = \hat{f} = \check{f} \}$ in the following sense:

THEOREM 8. Let $\{F_\beta\}$ be the Bishop decomposition for K. Then

$$A(K)|_{\overline{\partial K}} = \{ f \in C(\partial K) : f|_{\overline{\partial F_\alpha}} \in A(F_\alpha)|_{\overline{\partial F_\alpha}} \text{ and } f = \hat{f} = \check{f} \}$$

In [5] Ellis proved that if the Bishop decomposition for K covers $\overline{\partial K}$ then

$$A(K)|_{\overline{\partial K}} = \{ f \in C(\overline{\partial K}) : f|_{F_\beta \cap \overline{\partial K}} \in A(F_\beta)|_{F_\beta \cap \overline{\partial K}} \text{ for all } \beta \} \quad (*)$$

Now, if $f \in C(\overline{\partial K})$ and if $f|_{F_\beta \cap \overline{\partial K}} \in A(F_\beta)|_{F_\beta \cap \overline{\partial K}}$ then $f = \hat{f} = \check{f}$ on $F_\beta \cap \overline{\partial K}$ ([1], Corollary I.3.6.) so that Theorem 8 generalizes Ellis' result. As ([4], Example 10) shows, the set on the right hand side in (*) is in general larger than the set $A(K)|_{\overline{\partial K}}$. Now, it follows from Theorem 7 that any function in this larger set is annihilated by all boundary measures in $A(K)^{\perp}$. Thus, if the set on the right hand side in (*) is strictly larger than $A(K)|_{\overline{\partial K}}$, then it must contain functions f which fail to satisfy the requirement that $f = \hat{f} = \check{f}$. As it turns out we have:

THEOREM 9. *Let* $\{F_\beta\}$ *be the Bishop decomposition for* K. *Then*

$$\{ f \in C(\overline{\partial K}) : f|_{F_\beta \cap \overline{\partial K}} \in A(F_\beta)|_{F_\beta \cap \overline{\partial K}} \} =$$

$$\{ f \in C(\overline{\partial K}) : \int f d\mu = o \text{ for all } \mu \in A^{\perp} \cap M(\partial K) \text{ and }$$

$$f = \hat{f} = \check{f} \text{ on } \overline{\partial K} \cap (\cup F_\beta) \}$$

PROOF. It follows from the above remarks, that the set on the left is contained in the set on the right. Let f be a member of the set on the right. Since $\overline{\partial F_\beta} \subseteq \overline{\partial K}$ it follows from ([2] Theorem 1) that there is a function a in $A(F_\beta)$ such that $f|_{\overline{\partial F_\beta}} = a|_{\overline{\partial F_\beta}}$. We must show, that $f(x) = a(x)$ for all $x \in F_\beta \cap \overline{\partial K}$. Let $x \in F_\beta \cap \overline{\partial K}$. If $b \in A(K)$ and if $b|_{\overline{\partial K}} > f$ then $b|_{\overline{\partial F_\beta}} > a|_{\overline{\partial F_\beta}}$ and hence $\hat{f}(x) \geq a(x)$. Similarily $\check{f}(x) \leq a(x)$. But $f(x) = \hat{f}(x) = \check{f}(x)$ so that $f(x) = a(x)$.

REFERENCES.

1. E.M. Alfsen, Compact convex sets and boundary integrals (Springer-Verlag, Berlin, 1971).
2. E.M. Alfsen, 'On the Dirichlet problem of the Choquet boundary', Acta Math. 220 (1968), 149-159.
3. E.M. Alfsen and T.B. Andersen, 'Split faces of compact convex sets', Proc. London Math. Soc. 21 (1970), 415-442.
4. A.J. Ellis, 'Central decompositions and the essential set for the space A(K)', Proc. London Math. Soc. 26 (1973), 564-576.
5. A.J. Ellis, 'Central decompositions for compact convex sets', to appear.

BOUNDARY ZERO-SETS OF A^{∞} FUNCTIONS ON STRICTLY PSEUDO-CONVEX DOMAINS

by Anne-Marie Chollet

For certain algebras of analytic functions on strictly pseudo-convex domains in $\mathbb{C}^n$ sufficient conditions are given for a closed set on the boundary of the domain to be a zero-set.

Definitions and notations.

Let D be a bounded strictly pseudo convex domain in $\mathbb{C}^n$ with $C^{2+\varepsilon}$ boundary, $\varepsilon > 0$. Then there exists a $C^{2+\varepsilon}$ real function φ defined in a open neighborhood U of $\bar{D}$ such that

a) $D = \{z \in U \;\; ; \;\; \varphi(z) < 0\}$

b) φ is strictly plurisubharmonic on a neighborhood of ∂D

c) $\operatorname{grad} \varphi \neq 0$ on ∂D.

We denote by A(D) the class of functions analytic in D and continuous in $\bar{D}$, $A^m(D)$ [resp. $A^{\infty}(D)$] the class of functions analytic in D and continuous with all their derivatives up to order m [resp. of all orders] in $\bar{D}$.

For f belonging to $A^{\infty}(D)$ let

$Z^o(f) = \{ z \in \bar{D} \; ; \; f(z) = 0 \}$

$Z^m(f)$ the set of common zeros of f and all its derivatives up to order m, and $Z^\infty(f) = \bigcap_{m=0}^{\infty} Z^m(f)$.

<u>The case of the unit disc</u>.

Let $D = \Delta = \{ z \in \mathbb{C} \; ; \; |z| < 1 \}$.

It is well known that a necessary and sufficient condition for a closed set E on $\partial\Delta$ to be a zero-set for a function f belonging to $A(\Delta)$ is

$$\mu(E) = 0$$

if μ denotes the Lebesgue measure on $\partial\Delta$.

L. Carleson [1] has proved that a necessary and sufficient condition for a closed set E on $\partial\Delta$ to be a zero-set for a function belonging to $A^m(\Delta)$, $m \geq 1$ is

(C) $\qquad \mu(E) = 0 \quad \text{and} \quad \sum_\nu \ell_\nu \log 1/\ell_\nu < \infty$

if we denote by ℓ_ν the length of $]a_\nu, b_\nu[$, a complementary interval of E.

B. A. Taylor and D. L. Williams [6] and B. Korenblum [5] have independently proved that the same condition (C) is necessary and sufficient for

$$E = Z^o(f) = Z^\infty(f)$$

where f belongs to $A^\infty(\Delta)$.

<u>Pseudo-balls and pseudo-distance on</u> D.

With each point ζ of the boundary of D we associate ν_ζ, the unit outward normal and T_ζ the real tangent space at ζ. We denote by L_ζ the real subspace generated by $(i\nu_\zeta)$ and by N_ζ the complex subspace generated by ν_ζ.

We have the orthogonal sum decomposition

$$\mathbb{C}^n = N_\zeta \oplus P_\zeta \qquad \text{over } \mathbb{C}$$

and

$$T_\zeta = L_\zeta \oplus P_\zeta \qquad \text{over } \mathbb{R}.$$

Then P_ζ is the unique complex subspace of the real tangent space T_ζ with the complex dimension $n-1$.

We now define, on the boundary of D, a family of pseudo-balls $B(\zeta,r)$ of center ζ and radius r as follows

$$B(\zeta,r) = \left\{z \in \partial D \; ; \; |\langle z-\zeta, \nu_\zeta\rangle| < r \text{ and } |z-\zeta|^2 < r\right\}$$

where the symbol $\langle \; , \; \rangle$ denotes the hermitian product in $\mathbb{C}^n$.

In this way, we can define a pseudo-distance ρ on ∂D.

For all z and w, belonging to ∂D, $\rho(z,w) = \inf \{r$; there exists a pseudo-ball of radius r which contains z and $w\}$.

Then we have the following :

1) $\rho(z,w) = 0 \Longleftrightarrow z = w$

2) $\rho(z,w) = \rho(w,z)$ for each z and w in ∂D

3) There exists a constant K such that for all z, w and t in ∂D $\rho(z,w) \leq K\left[\rho(z,t) + \rho(t,w)\right]$.

Let us now use the abreviations balls and distance for pseudo-balls and pseudo-distance. With this notation, we can rewrite a theorem of A. M. Davie and B. K. Øksendal [4].

THEOREM 1. Let E be a closed subset of ∂D. A sufficient condition for E to be a zero-set for $A(D)$ is that, for every $\varepsilon > 0$, the set E can be covered by a sequence $\{B_i\}$ of open balls of radius r_i such that

$$\sum_i r_i < \varepsilon .$$

More precisely, A. M. Davie and B. K. Øksendal construct a function f of $A(D)$ with $f = 0$ on E and $\operatorname{Re} f > 0$ on $\bar{D} \setminus E$ such that $g = \exp(-f)$ peaks on E. In $A(D)$, the notions of zero set, peak set and peak interpolation set are equivalent. This is no longer true in $A^m(D)$, $m > 1$. Here, as in the case of the disk [7], the only peak sets are finite sets.

Whereas the proof of theorem 1 used an open covering of E, in the case of $A^\infty(D)$, as in [5], we use an open covering of $\complement E$, the complement of E.

We need the two following lemmas.

LEMMA 1 [3]. There exist constants k, A, B and M. such that if E is a closed subset of ∂D, there is a sequence of balls $\{B(\zeta_i, r_i)\}$ satisfying

1) the balls $B(\zeta_i, r_i)$ are disjoint.

2) $\complement E \subset \bigcup_{i=1}^{\infty} B(\zeta_i, kr_i)$.

3) If z belongs to a ball $B(\zeta_i, kr_i)$ we have

$$Ar_i \le \rho(z, E) \le Br_i .$$

4) A point z of the complement of E cannot belong to more than M distinct balls.

DEFINITION. If a sequence of balls $\{B(\zeta_i, r_i)\}$ satisfies the properties 1) to 4), the set of dilated balls $B(\zeta_i, kr_i)$ will be called a Whitney covering of $\complement E$.

LEMMA 2. Let E be a closed subset of ∂D in $\mathbb{C}^n$ and $\{B(\zeta_i, r_i)\}$ a sequence of balls for which the set of dilated balls $B(\zeta_i, kr_i)$ is a Whitney covering of $\complement E$. Then the following conditions are equivalent :

(1) $$\int_{\partial D} \frac{1}{\left[\rho(z,E)\right]^{n-1}} \log \frac{1}{\rho(z,E)} \, d\mu(z) < \infty$$

where μ denotes the surface measure on ∂D.

(2) $$\mu(E) = 0 \quad \text{and} \quad \sum_i r_i \log \frac{1}{r_i} < \infty.$$

Remark. As in the case of the disc the distance $\rho(z,w)$ reduces to the euclidean distance we see that these conditions extend the condition (C).

THEOREM 2. Let E be a closed subset of ∂D satisfying one of the two equivalent conditions of lemma 2. Then, there exists a function F belonging to $A^\infty(D)$ such that

$$E = Z^0(F) = Z^\infty(F).$$

Sketch of the proof. For any $\delta > 0$, we write $D_\delta = \{z \in \mathbb{C}^n \,;\, \varphi(z) < \delta\}$ and we define $\mathcal{C}(\partial D, \mathcal{H}(D_\delta))$ to be the space of functions continuous on ∂D with values in $\mathcal{H}(D_\delta)$, the space of functions holomorphic in D_δ.

Then there exist constants M, m, δ, strictly positive, and a function G such that

a) G belongs to $\mathcal{C}(\partial D, \mathcal{H}(D_\delta))$

b) For each (ζ, z) in $\partial D \times \bar{D}$

$$\operatorname{Re} G(\zeta, z) \leq -m\,|\zeta - z|^2$$

c) For each (ζ,z) in $\partial D \times \partial D$

$$\operatorname{Re} G(\zeta,z) \geq -M|\zeta - z|^2$$

d) For each ζ in ∂D

$$\operatorname{grad}_z \operatorname{Re} G(\zeta,\zeta) = \frac{1}{2} \operatorname{grad} \varphi(\zeta).$$

Such a function G has already been used by A. M. Davie and B. K. Øksendal in [4], but here as we use an open covering of the complement of E we need constants m and M independent of the choice of ζ on ∂D.

Moreover, it is possible to choose G so that there exists constants A and B, strictly positive, such that, for each (ζ,z) in $\partial D \times \partial D$

$$B\rho(\zeta,z) \leq |G(\zeta,z)| \leq A\rho(\zeta,z).$$

In the case of the unit ball in $\mathbb{C}^n$, $D = \{z \in \mathbb{C}^n \; ; \; |z|^2 - 1 < 0\}$, then we have $\rho(\zeta,z) = |1 - \langle \zeta,z \rangle|$ and we can take

$$G(\zeta,z) = \langle \zeta,z \rangle - 1$$

Let E be a closed set E on ∂D satisfying the hypothesis of the theorem :

$$\mu(E) = 0 \quad \text{and} \quad \sum_i r_i \log \frac{1}{r_i} < \infty.$$

Let λ_i be a sequence of real numbers tending to infinity such that $\sum_i \lambda_i r_i \log \frac{1}{r_i} < \infty$.

Then we can prove that

$$F(z) = \exp \sum_i \frac{\lambda_i r_i \log 1/r_i}{G(\zeta_i,z) - r_i}$$

belongs to $A^\infty(D)$ and satisfies

$$E = Z^0(F) = Z^\infty(F).$$

COROLLARY. Let Γ be an arc in ∂D of class $C^{1+\alpha}$, $\alpha > 0$, whose tangent at each point ζ lies in the complex tangent subspace P_ζ. Then there exists f belonging to $A^\infty(D)$ such that

$$E = Z^0(f) = Z^\infty(f).$$

Such a set E has already been proved in [4] to be a zero-set for $A(D)$.

Remark. In the same way, we can get sufficient conditions for more special classes of functions of $A^\infty(D)$, for instance, for Gevrey classes. These results extend those obtained in [2], in the case of the unit disc.

Bibliography

[1] CARLESON, L. Sets of uniqueness for functions regular in the unit circle. Acta Math. 87 (1952), 325-345.

[2] CHOLLET, A.-M. Ensembles de zéros de fonctions analytiques dans le disque. C. R. Acad. Sc. Paris 276 (1973), 731-733.

[3] COIFMAN, R. and WEISS, G. Analyse harmonique non commutative sur certains espaces homogènes. Springer Verlag, 1971.

[4] DAVIE, A. M. and ØKSENDAL, B. K. Peak interpolation sets for some algebras of analytic functions. Pacific J. Math. 41 (1972), 81-87.

[5] KORENBLUM, B. Functions holomorphic in a disk and smooth in its closure. Soviet Math. Dokl. 12 (1971), 1312-1315.

[6] TAYLOR, B. A. and WILLIAMS, D. L. Ideal in rings of analytic functions with smooth boundary values. Can. J. Math. 22 (1970), 1266-1283.

[7] TAYLOR, B. A. and WILLIAMS, D. L. The peak sets of A^m. Proc. Amer. Math. Soc. 24 (1970), 604-605.

HIGHER POINT DERIVATIONS ON COMMUTATIVE BANACH ALGEBRAS

H.G. Dales

1. This paper is a summary of part of some joint work with J.P. McClure (Winnipeg, Canada) now in preparation ([3], [4]).

Let A be a commutative Banach algebra and let d_o be a character on A. A point derivation of order q (respectively, of infinite order) on A at d_o is a sequence of linear functionals on A such that, for f and g in A and $k = 1,\ldots,q$ (respectively, $1,2,\ldots$),

$$d_k(fg) = \sum_{i=0}^{k} d_i(f)d_{k-i}(g).$$

These equations are called the Leibnitz identities. Of course, d_1 is a point derivation at d_o in the usual sense. The point derivation is continuous if d_i is continuous for each $i \geq 1$, and it is totally discontinuous if d_i is discontinuous for each $i \geq 1$.

Example: Let A be the standard disc algebra and let $d_k(f) = f^{(k)}(0)/k!$ $(f \in A)$. This gives a continuous point derivation of infinite order on A at the origin.

We say that a point derivation $d_1,\ldots,d_q$ belongs to a point derivation of order p (where $p > q$) if there are linear functionals $d_{q+1},\ldots,d_p$ such that $d_1,\ldots,d_p$ is a point derivation of order p.

A point derivation is non-degenerate if $d_1 \neq 0$.

We wish to investigate these (higher) point derivations on particular commutative Banach algebras, and we ask the questions:

Question I. Is there a function $q \mapsto p(q)$ on the positive integers such that, whenever a point derivation of order q belongs to a point derivation of order $p(q)$, the point derivation of order q is necessarily continuous?

For many of the familiar Banach algebras known to have discontinuous point derivations, the answer is ''yes'', with $p(q)$ no larger than $2q$.

Question II. Given k, what is the maximum order of a point derivation satisfying the condition that $d_k \neq 0$?

Again, for many Banach algebras, the maximum order is finite and can be determined.

2. The most interesting example of a Banach algebra with discontinuous point derivations is $C^{(n)} = C^{(n)}(I)$, the Banach algebra of n-times continuously differentiable functions on the interval $I = [0,1]$. The theory of derivations from this algebra into certain modules has been studied by Badé and Curtis, [1].

We can show for point derivations on $C^{(n)}$ at the point 0 of I the following results.

(i) A continuous, non-degenerate point derivation has order at most n.

(ii) The maximum order asked for in question II, above, is exactly $(2n+1)k - 1$, so that a non-degenerate point derivation has order at most $2n$.

(iii) Let $d_1,\ldots,d_p$ be a point derivation of order p. If $p \geq 2q$, then the point derivation $d_1,\ldots,d_q$ is continuous.

(iv) A continuous, non-degenerate point derivation $d_1,\ldots,d_q$ of order $q < n$ belongs to a point derivation of order $2q + 1$ with d_j discontinuous for $j = q + 1,\ldots,2q + 1$.

(v) A continuous, non-degenerate point derivation of order n belongs to one of order $2n$ with d_j discontinuous for $j = n + 1,\ldots,2n$.

Thus, we see from (iii) that the function $p(q) = 2q$ answers question I. It is interesting that this function is independent of n, and also that we do not have to assume that the point derivation is non-degenerate. That the bound $2q$ is best-possible is shown by (iv) and (v).

An important technical result for the above algebras is the following, proved by both A.Browder and P.C.Curtis. Let

$$M_{nk} = \{f \in C^{(n)} : f(0) = \dots = f^{(k)}(0) = 0\}.$$

Then $M_{nn}{}^2 = x^n M_{nn}$ and, for $0 \leq k < n$, $M_{nk}{}^2 = x^{k+1}M_{nk}$.

Typically, our proofs of the above results use the Browder-Curtis Lemma and a lot of induction.

For other examples, all with $p(q) = 2q$, see [3].

3. Could it be that a function $p(q)$ exists for every Banach algebra with discontinuous first order point derivations, perhaps with $p(q) = 2q$? We consider the construction of totally discontinuous point derivations which will provide counter-examples to this possibility.

Let $M = \ker d_o$ be a maximal ideal of an algebra A. Of course, first order point derivations can be identified with linear functionals d on M such that $d \perp M^2$. So, if M^2 has infinite codimension in M, we can obtain (but not 'construct') a discontinuous d.

Now suppose that we wish to obtain d_o, d, d_2, a totally discontinuous point derivation of order two. Starting with d as above, we must define d_2 on M^2 by

$$d_2\left(\sum_{i=1}^{n} f_i g_i\right) = \sum_{i=1}^{n} d(f_i)d(g_i),$$

and extend d_2 arbitrarily (linear) to M. Everything works, provided d_2 is well-defined on M^2. So the question is : can we find a maximal ideal M in a Banach algebra and a linear functional d on M such that $d \perp M^2$, $\sum_{i=1}^{n} f_i g_i = 0 \Rightarrow \sum_{i=1}^{n} d(f_i)d(g_i) = 0$, and d is discontinuous ?

In fact, we can do this and considerably more.

4. Let $\mathcal{F} = \mathbb{C}[[X]]$ be the algebra of formal power series in one variable over $\mathbb{C}$, and let $p_j : \Sigma \lambda_i X^i \to \lambda_j$ be the coordinate projections on $\mathcal{F}$. Then $\mathcal{F}$ is a complete Fréchet algebra with respect to the topology determined by the semi-norms $|p_j|$. Clearly, if $\{d_i\}$ is a point derivation of infinite order on A, then $a \to \Sigma d_i(a)X^i$ is a homomorphism $A \to \mathcal{F}$. (So, for example, we see from 2(ii) that every homomorphism from $C^{(n)}(I)$ into $\mathcal{F}$ has the form $f \to f(t_0)1$, some $t_0 \in I$, where 1 is the ring identity of $\mathcal{F}$. In particular, every such homomorphism is continuous.) Can there be a discontinuous homomorphism from a Banach algebra into $\mathcal{F}$? A totally discontinuous point derivation of infinite order would give such a homomorphism.

The possibility that every homomorphism into $\mathcal{F}$ is continuous is suggested by the following result, [8],[11]. A subalgebra B of $\mathcal{F}$ is a <u>Banach algebra of power series</u> if it contains the polynomials and is a Banach algebra under a norm such that the inclusion map is continuous – see [5]. Then every homomorphism from a Banach algebra into B is necessarily continuous. The result extends to certain Fréchet algebras of power series, but not to $\mathcal{F}$ itself.

Can there be a homomorphism from a Banach algebra <u>onto</u> $\mathcal{F}$? It is easy to see that such an epimorphism must be discontinuous.

We can answer these various questions by proving the following theorem.

<u>Theorem</u>. There is a Banach algebra A and a totally discontinuous point derivation of infinite order $\{d_i\}$ at the character d_0 of A such that the map $a \to \Sigma d_i(a)X^i$, $A \to \mathcal{F}$, is an epimorphism.

5. We give a brief sketch of the construction of the algebra A. In its present formulation, the example is due to Peter McClure.

Let $B = B(1)$ be an infinite-dimensional Banach space, take $B(0) = \mathbb{C}$, and, for $k = 2,3,\ldots$, let $B(k)$ be the completion of $B^{\otimes k}$ with respect to the ε-norm (this is also called the weak tensor norm). Let

$$\widehat{\widehat{\otimes}}\, B = \{ \sum_0^\infty u_k : u_k \in B(k) \quad (k = 0,1,2,\ldots)\}.$$

With coordinatewise addition and scalar multiplication and the topology of coordinatewise convergence, $\widehat{\widehat{\otimes}}\, B$ is a Fréchet space.

Let $\otimes B$ be the usual (algebraic) tensor algebra over B with multiplication $\otimes$ (e.g., [6, §3.2]), so that $\otimes B$ is an associative algebra with identity. Thus, if $u = \Sigma u_k$, $v = \Sigma v_k$ in $\otimes B$, $u \otimes v = \Sigma_k (\Sigma_{i+j=k} u_i \otimes v_j)$. If we extend the multiplication to $\widehat{\widehat{\otimes}}\, B$, it is easy to see that we get a Fréchet algebra with identity. (We might term $\widehat{\widehat{\otimes}}\, B$ a 'topological graded algebra'.)

The algebra $\widehat{\widehat{\otimes}}\, B$ is very non-commutative, and, for our purposes, we need a commutative version of the above construction. For each k, let $\mathcal{S}_k$ denote the symmetric group on k symbols, and for $\sigma \in \mathcal{S}_k$, $b_1,\ldots,b_k \in B$, define $\sigma(b_1 \otimes \ldots \otimes b_k) = b_{\sigma(1)} \otimes \ldots \otimes b_{\sigma(k)}$. Then σ extends to a well-defined linear isometry on $B^{\otimes k}$ and hence on $B(k)$. Let $S_k = \frac{1}{k!}\{\Sigma\, \sigma : \sigma \in \mathcal{S}_k\}$ be the symmetrizing map, and let $SB(k)$ denote the symmetric elements in $B(k)$, so that $SB(k) = \{u : \sigma u = u$ for each $\sigma \in \mathcal{S}_k\}$. Now let

$$\widetilde{\vee}\, B = \{ \sum_0^\infty u_k : u_k \in SB(k) \quad (k = 0,1,2,\ldots)\},$$

a closed linear subspace of $\widehat{\widehat{\otimes}}\, B$. With ring multiplication based on the symmetric tensor product, $u \vee v = S_{p+q}(u \otimes v)$ for $u \in SB(p)$, $v \in SB(q)$, $\widetilde{\vee}\, B$ is a commutative topological graded Fréchet algebra with identity. In fact, the algebra $\widetilde{\vee}\, B$ which we have described is a certain completion of $\vee B$, the standard symmetric algebra over B ([6, §7.3]).

To obtain the algebra of the theorem, let $\{\lambda_k\}$ be any sequence of positive numbers such that $\lambda_o = 1$ and $\lambda_{p+q} \leq \lambda_p \lambda_q$ for each p and q. Let

$$A = \{u = \Sigma\, u_k \in \tilde{\bigvee} B : \|u\| = \sum_{k=o}^{\infty} \|u_k\| \lambda_k < \infty\}.$$

It is easy to check that A is a commutative Banach algebra with identity.

Write P_j for the natural projection map $\Sigma\, u_k \to u_j$, $\tilde{\bigvee} B \to SB(j)$. Then P_o is a character on $\tilde{\bigvee} B$, and the sequence of maps $\{P_j\}$ satisfies the equations

$$P_k(u \vee v) = \sum_{j=o}^{k} P_j(u) \vee P_{k-j}(v).$$

Let Λ_o be the identity map on $\mathbb{C}$ and let Λ_j be a linear functional on SB(j) for $j = 1,2,\ldots$. Define $d_j = \Lambda_j \circ P_j$. Then $\{d_i\}$ will be a point derivation at d_o if and only if

$$\Lambda_{p+q}(u \vee v) = \Lambda_p(u)\Lambda_q(v)$$

for any p and q and any $u \in SB(p)$, $v \in SB(q)$. Given Λ_1 a <u>continuous</u> linear functional on B, it is easy to see that suitable $\Lambda_2, \Lambda_3, \ldots$ exist (all continuous). What is perhaps surprising is that $\Lambda_2, \Lambda_3, \ldots$ can be found for any Λ_1, even discontinuous. The proof of this is rather long, and is the crucial part of the theorem.

Using the discontinuity of Λ_1 and the completeness of A, it is not hard to check that, given an element of $\mathcal{F}$, we can find a preimage in A, so that we have constucted the required epimorphism.

REFERENCES

We include certain related papers in this list of references.

[1] W.G.Badé and P.C.Curtis, Jr., The continuity of derivations of Banach algebras, J. Functional Analysis 16 (1974), 372-387.

[2] H.G.Dales and J.P.McClure, Continuity of homomorphisms into certain commutative Banach algebras, Proc. London Math. Soc. (3) 26 (1973), 69-81.

[3] ——————, Higher point derivations on commutative Banach algebras, I, preprint.

[4] ——————, ——————, II and III, in preparation.

[5] S.Grabiner, 'Derivations and automorphisms of Banach algebras of power series' Mem. American Math. Soc., no. 146, A.M.S., 1974.

[6] W.H.Greub, 'Multilinear algebra', Springer-Verlag, 1967.

[7] F.Gulick, Systems of derivations, Trans. American Math. Soc. 149 (1970), 465-488.

[8] B.E.Johnson, Continuity of linear operators commuting with continuous linear operators, Trans. American Math. Soc. 128 (1967), 88-102.

[9] R.J.Loy, Continuity of higher derivations, Proc. American Math. Soc. 37 (1973), 505-510.

[10] ———, Commutative Banach algebras with non-unique complete norm topology, Bull. Australian Math. Soc. 10 (1974), 409-420.

[11] ———, Banach algebras of power series, J. Australian Math. Soc. 17 (1974), 263-273.

CLASSIFICATION OF ESSENTIALLY NORMAL OPERATORS

A.M. DAVIE

This is an attempt to expound part of a theory recently developed by Brown, Douglas and Fillmore [3,4]. They consider the following problem: classify the normal elements of the Calkin algebra $\mathcal{A}$ (i.e. the quotient of the algebra of all bounded linear operators on Hilbert space by the ideal of compact operators) up to unitary equivalence. Two unitary invariants for such elements are easily described: the spectrum (a compact subset of $\mathbb{C}$) and the Fredholm index (for details see Section 1 below). One of the main results of [4] is that these two invariants are sufficient for a complete unitary classification of the normal elements of $\mathcal{A}$. As a corollary, the set of operators which can be expressed as the sum of a normal operator and a compact operator, is (norm) closed. Our first objective is to give a somewhat simplified version of the proof of these results.

The strategy of the proof, both in [4] and here, is to associate to each compact subset X of $\mathbb{C}$, the set of all unitary equivalence classes of normal elements with spectrum X, denoted $\mathrm{Ext}(X)$, and to show essentially that $\mathrm{Ext}(X)$ behaves like a homology theory. One can then use techniques of algebraic topology to determine $\mathrm{Ext}(X)$, which is equivalent to solving the original problem. In [4] the operator theory and algebraic topology are interwoven throughout much of the proof. Using a simplification of part of the proof due to Arveson, we are able to separate the operator theory from the algebraic topology. In section 3 we list seven properties of $\mathrm{Ext}(X)$, which may be thought of as "axioms" for a homology theory. These are proved in section 4, by operator theoretic methods (some simplifications over [4] are achieved by restricting to the case $X \subseteq \mathbb{C}$). Then in section 5 we show how the seven properties determine $\mathrm{Ext}(X)$, giving the result quoted above. This is purely an exercise in algebraic topology. (It is quite elementary, and assumes no previous knowledge of algebraic topology.) It is hoped that this approach will make the proof understandable without any background in algebraic topology.

Nevertheless, our second objective is to convince the reader that the algebraic topology is an essential part of the problem. This comes out most clearly from the theory outlined in [3] of $\mathrm{Ext}(X)$ for general compact metric X. For $X \subseteq \mathbb{C}^n$ this is the problem of classi-

fying n-tuples of commuting normal elements of $\mathcal{a}$. The general
theory involves sophicticated algebraic topology (mainly K-theory).
One can however, treat some illuminating special cases on the basis
of the seven properties, and this we do in section 6. (We cheat
slightly, because we prove the seven properties only if $X \subseteq \underset{\sim}{C}$ - this
does not, however, evade any serious difficulties.) One of the most
striking results is as follows: consider the set of n-tuples of self-
adjoint operators which can be expressed in the form $(T_1 + K_1, \ldots, T_n + K_n)$
where $T_1 \ldots T_n$ are commuting self-adjoint operators and $K_1 \ldots K_n$ are
compact self-adjoint operators. Question: is this set (norm)   closed?
(The case n = 2 is the result quoted above.) Answer: yes, if $n \leq 3$,
no, if $n \geq 4$. An explicit example for n = 4 is given in section 6;
what makes it possible, is a certain topological phenomenon which can
occur in $\underset{\sim}{R}^4$ but not in $\underset{\sim}{R}^3$.

Here we have two quite dissimilar fields of mathematics - operator
theory and algebraic topology - interacting in a deep, essential and
unexpected manner. In my view it is this, much more than the solution
of an interesting and difficult problem in operator theory, which
makes the work of Brown, Douglas and Fillmore an outstanding piece of
mathematics.

1. Operator Theory Background

In this section we fill in the operator-theoretic background to
the problem, and also formulate the main theorem precisely.

We shall be working throughout on a complex separable infinite-
dimensional Hilbert space H . The term operator will always mean
bounded linear operator on H . B(H) denotes the algebra of all such
operators.

For orientation and future reference, we mention the classification of normal operators up to unitary equivalence, which follows from
spectral theory:- given a normal operator N , we can find positive
measures $\mu_1, \mu_2, \ldots$ on a bounded subset of $\underset{\sim}{C}$, with $\mu_{n+1} \ll \mu_n$,
such that N is unitarily equivalent to $M_1 \oplus M_2 \oplus \ldots$ where M_n
is multiplication by z on $L^2(\mu_n)$. The measures μ_n are uniquely
determined up to mutual absolute continuity. See [7,X.5.10].

Our real concern is classification modulo the compact operators,
or classification of elements of the Calkin algebra $\mathcal{a}$. Given two
operators S and T , we write $S \sim T$ if there exist a compact operator K and a unitary operator U such that $S = U^{-1}TU + K$. If
$\pi : B(H) \rightarrow \mathcal{a}$ is the quotient mapping ($\mathcal{a}$ being the quotient of

B(H) by the ideal of compact operators) then $S \sim T \iff$ there exists a unitary operator U such that $\pi(S) = \pi(U)^{-1}\pi(T)\pi(U)$. It is clear that one invariant of the relation is the spectrum $\sigma(\pi(T))$ of $\pi(T)$. This is also known as the essential spectrum of T and denoted by $\sigma_e(T)$. In 1971 Berg [2] proved that if S and T are normal and $\sigma_e(T) = \sigma_e(S)$, then $S \sim T$ (following earlier work of Weyl and von Neumann). Thus the classification of normal operators up to $\sim$ is simpler than the unitary classification. We shall prove Berg's result later.

The class of operators we shall be concerned with, is that of essentially normal operators, i.e. operators T such that $T^*T - TT^*$ is compact, or equivalently $\pi(T)$ is a normal element of $\mathcal{A}$. The essential spectrum is no longer adequate to classify these operators up to $\sim$, as we shall see in a moment, and we have to introduce another invariant, the Fredholm index.

For any operator T , $\pi(T)$ is invertible in $\mathcal{A}$ if and only if T is Fredholm, i.e. T has closed range, and its kernel and cokernel are both finite-dimensional. If this is the case, one can define an integer $\operatorname{ind}(T) = \dim \ker T - \dim \operatorname{coker} T$. Then $\operatorname{ind}(T+K) = \operatorname{ind}(T)$ whenever K is compact, so $\operatorname{ind}(T)$ depends only on $\pi(T)$. Also $\operatorname{ind}(ST) = \operatorname{ind}(S) + \operatorname{ind}(T)$, so ind defines a homomorphism from the group $\mathcal{A}^{-1}$ of invertible elements of $\mathcal{A}$ to the group $\mathbb{Z}$ of integers. Moreover, this homomorphism is continuous, so ind is constant on each component of $\mathcal{A}^{-1}$. For details, see Chapter 5 of [6].

If S and $T \in B(H)$ and $S \sim T$, then we have $\operatorname{ind}(S-\lambda) = \operatorname{ind}(T-\lambda)$ for all $\lambda \notin \sigma_e(S)$. For any $T \in B(H)$, $\operatorname{ind}(T-\lambda)$ is constant on each component of $\mathbb{C} \setminus \sigma_e(T)$, and zero on the unbounded one (because $T-\lambda$ is invertible if $|\lambda|$ is large enough). Thus ind assigns an integer to each bounded component of $\mathbb{C} \setminus \sigma_e(T)$. We can now state the main result:

<u>Theorem</u>. Let S and T be essentially normal. Then $S \sim T$ if and only if $\sigma_e(S) = \sigma_e(T)$ and $\operatorname{ind}(S-\lambda) = \operatorname{ind}(T-\lambda)$ for all $\lambda \in \mathbb{C}\setminus\sigma_e(S)$.

For normal operators, the index is always zero, so the above theorem contains the Berg result. Taking the case where S is normal, we obtain the following Corollary.

<u>Corollary 1</u>: Let $T \in B(H)$. Then T is of the form $N + K$, where N is normal and K is compact, if and only if T is essentially normal and $\operatorname{ind}(T-\lambda) = 0$ for all $\lambda \in \mathbb{C}\setminus\sigma_e(T)$.

In particular, any essentially normal operator whose essential spectrum has connected complement, is of the form $N+K$.

Corollary 2: The set of operators of the form $N+K$ (N normal, K compact) is (norm) closed.

As an example consider the unilateral shift S, given by $Se_n = e_{n+1}$ for some orthonormal basis $(e_0, e_1, \ldots)$. Then $S^*S = 1$, and $SS^* = 1 - P$ where P is the projection on the 1-dimensional space spanned by e_0. So $\pi(S)$ is unitary, so $\sigma_e(S) \subseteq \Gamma$, the unit circle. Also $\ker S = 0$, whereas $\ker S^*$ is 1-dimensional. So $\operatorname{ind}(S) = -1$. So S is not of the form $N+K$. (It also follows that 0 is not in the unbounded component of $\mathbb{C} \setminus \sigma_e(S)$, so $\sigma_e(S)$ is the whole circle.)

If, however, T is a normal operator whose essential spectrum is the closed unit disc, then Cor. 1 shows that the direct sum $S \oplus T$ is of the form $N+K$. This fact was proved directly in 1972 by Deddens and Stampfli [5].

We conclude this section with an observation which forms the basis of the topological approach to the proof. Since $\mathcal{A}$ is a C*-algebra, if α is a normal element of $\mathcal{A}$, then the closed subalgebra of $\mathcal{A}$ generated by α and α^* is a commutative C*-algebra, and so is *-isomorphic to $C(\sigma(\alpha))$, and under this isomorphism α corresponds to the function $\varphi(z) = z$. So we can define $f(\alpha)$ for any $f \in C(\sigma(\alpha))$. This allows us to widen the definition of index - for $f \in C(\sigma(\alpha))^{-1}$, $f(\alpha) \in \mathcal{A}^{-1}$, so we can define $\operatorname{ind} f(\alpha)$. This gives a homomorphism from the group $C(\sigma(\alpha))^{-1}$ to $\mathbb{Z}$, which is continuous and hence constant on each component of $C(\sigma(\alpha))^{-1}$.

2. Topological Background

We collect here a number of topological results and concepts which will be used later.

Let X be a compact metric space, and let $B(X)$ denote the quotient of the multiplicative group $C(X)^{-1}$ of continuous non-vanishing functions on X, by its component of the identity, which is a clopen subgroup. It is an elementary fact (valid in any commutative Banach algebra) that the component of the identity in $C(X)^{-1}$ is the set of functions e^f where $f \in C(X)$. ($B(X)$ is known variously as the Bruschlinsky group, first integral Čech cohomology group, and first cohomotopy group of X. In [4] it is denoted $\pi^1(X)$.)

In the case where $X \subseteq \mathbb{C}$ we need the following description of $B(X)$. If $\lambda \in \mathbb{C}$, φ_λ will denote the function $\varphi_\lambda(z) = z - \lambda$.

<u>Proposition 2.1</u> Let X be a compact subset of $\underset{\sim}{C}$. Let Λ be a set consisting of one point from each bounded component of $\underset{\sim}{C}\setminus X$. Then $B(X)$ is freely generated as an Abelian group by $\{[\varphi_\lambda] : \lambda \in \Lambda\}$. (Here $[f]$ is the coset of f in $B(X)$.)

In other words, each element of $B(X)$ is uniquely expressible as $[\varphi_{\lambda_1}^{k_1} \dots \varphi_{\lambda_n}^{k_n}]$ where $\lambda_1 \dots \lambda_n \in \Lambda$, $k_1 \dots k_n \in \mathbb{Z}$.

<u>Sketch of proof</u>. This is fairly straightforward in the special case X is a finite union of disjoint compact sets, each bounded by finitely many disjoint smooth Jordan curves. In this case $f \in C(X)^{-1}$ is of the form e^g $(g \in C(X))$ if and only if the variation of arg. f around each of these curves is zero. Given f we can find a product of φ_λ's with the same variation of argument around each curve as f, so the $\{\varphi_\lambda\}$ generate $B(X)$. Also any non-trivial product of φ_λ's has non-zero variation around some curve, so gives a non-trivial element of $B(X)$.

The general case can be reduced to this special case: let X_n be a decreasing sequence of sets of the type considered above, with intersection X. Then if $f \in C(X)^{-1}$ we can extend f to a continuous function on $\underset{\sim}{C}$, and then f will be non-zero on X_n for some n. Then we have $f = e^g h$ on X_n where $g \in C(X_n)$, and h is a product of φ_μ's for $\mu \in \underset{\sim}{C}\setminus X_n$. But for each such μ, μ is in the same component of $\underset{\sim}{C}\setminus X$ as some $\lambda \in \Lambda$ and it is elementary that then $\varphi_\mu^{-1}\varphi_\lambda$ has a logarithm in $C(X)$. This shows that the φ_λ generate $B(X)$. A similar reduction to the special case shows that they generate freely.

We denote by $G(X)$ the group of all group homomorphisms from $B(X)$ to $\mathbb{Z}$, or equivalently the group of all continuous group homomorphisms from $C(X)^{-1}$ to $\mathbb{Z}$. Proposition 2.1 then identifies $G(X)$ with the group of all ways of assigning an integer to each bounded component of $\underset{\sim}{C}\setminus X$, when $X \subseteq \underset{\sim}{C}$. If α is a normal element of $\mathcal{A}$ with spectrum X, then the observation at the end of section 2 associates to α an element of $G(X)$ which we denote by $i(\alpha)$. Then $i(\alpha)$ is determined by $i(\alpha)[\varphi_\lambda] = \operatorname{ind}(\varphi_\lambda(\alpha)) = \operatorname{ind}(\alpha-\lambda)$ for $\lambda \in \underset{\sim}{C}\setminus X$.

In [4] $G(X)$ is denoted by $\operatorname{Hom}(\pi^1(X),\mathbb{Z})$.

We note that a continuous map $f : X \to Y$ induces naturally a homomorphism $f_+ : G(X) \to G(Y)$.

We conclude with a few notions that will be needed later.

(a) If X is compact metric and $A \subseteq X$ is closed, we denote by X/A the compact space obtained from X by identifying A to a point. There is then a natural quotient map $p : X \to X/A$.

(b) If we are given compact metric spaces $X_1, X_2, \ldots$ and continuous maps $\rho_n : X_{n+1} \to X_n$ then we define the projective limit X of the system (X_n, ρ_n) as the set of all sequences $x = (x_1, x_2, \ldots)$ with $x_n \in X_n$ and $\rho_n x_{n+1} = x_n$, with the product topology. X is compact metric and there are natural maps $\pi_n : X \to X_n$ given by $\pi_n x = x_n$. The algebra of continuous functions on X which factor through π_n for some n separates points of X, so is dense in $C(X)$.

(c) The cone on X, denoted by $\text{cone}(X)$, is defined as $[0,1] \times X / \{0\} \times X$. The suspension of X, denoted by SX, is obtained from $[0,1] \times X$ by identifying $\{0\} \times X$ and $\{1\} \times X$ to (different) points. One can think of SX as $\text{cone}(X) / \{1\} \times X$.

(d) A closed subset A of X is a retract of X if there is a continuous map $r : X \to A$ with $r(a) = a$, $a \in A$.

3. Definition of Ext(X) and statement of properties

We give 3 equivalent formulations of the definition of $\text{Ext}(X)$, where $X \subseteq \underset{\sim}{C}$, labelled (A), (B), (C).

(A) $\text{Ext}(X)$ is the set of all equivalence classes under $\sim$ of essentially normal operators T with $\sigma_e(T) = X$. We make $\text{Ext}(X)$ into an Abelian group by putting $[S] + [T] = [S \oplus T]$. (It is easy to check that the equivalence class of $S \oplus T$ depends only on the classes of S and T, and is the same as that of $T \oplus S$.)

(B) $\text{Ext}(X)$ is the set of equivalence classes of normal elements of $\mathcal{Q}$ with spectrum X, where we define α and β to be equivalent if $\alpha = \pi(U)^{-1}\beta\pi(U)$ for some unitary $U \in B(H)$. Again we define the semigroup operation by direct sums.

(C) $\text{Ext}(X)$ is the set of equivalence classes of $(1-1)$ *-homomorphisms $\lambda : C(X) \to \mathcal{Q}$, where λ_1 and λ_2 are defined to be equivalent if there is a unitary $U \in B(H)$ such that $\lambda_1(f) = \pi(U)^{-1}\lambda_2(f)\pi(U)$ for $f \in C(X)$. Again we define the semi-group operation by direct sums - $[\lambda_1] + [\lambda_2] = [\lambda]$ where $\lambda(f) = \lambda_1(f) \oplus \lambda_2(f)$, regarded as an element of the Calkin algebra on $H \oplus H$.

The equivalence of (A) and (B) is clear; the equivalence of (B) and (C) follows from the last paragraph of section 1 - given α, normal in $\mathcal{A}$ with spectrum X, define $\lambda : C(X) \to \mathcal{A}$ by $\lambda(f) = f(\alpha)$. Conversely given λ, we recover α by $\alpha = \lambda(\varphi)$ where $\varphi(z) = z$.

The problem of classifying the essentially normal operators up to $\sim$ is equivalent to determining $\mathrm{Ext}(X)$ for all compact $X \subseteq \underset{\sim}{C}$. Before listing the properties of Ext which will enable us to do this, we make some remarks:

(i) As an example of formulation (C), let S be the unilateral shift and consider its equivalence class in $\mathrm{Ext}(\Gamma)$. The associated $\lambda : C(\Gamma) \to \mathcal{A}$ is given by $\lambda(f) = f(\pi(S))$. Now S can be thought of as multiplication by z on the Hardy space H^2. Then one can show that $\lambda(f) = \pi(T_f)$ where T_f is the Toeplitz operator with symbol f.

(ii) (C) makes sense for any compact metric X, so can be used as a definition of $\mathrm{Ext}(X)$ in this generality. See section 6.

(iii) In formulations (B) and (C), it may be asked whether it would be more natural to replace $\pi(U)$ by a unitary element of $\mathcal{A}$, thus giving an apparently weaker notion of equivalence. In fact, it is shown in [4, Theorem 4.3] that this notion is the same as ours.

<u>Properties of Ext(X)</u>

(1) $\mathrm{Ext}(X)$ is an Abelian group.

(2) A continuous map $f : X \to Y$ induces a homomorphism $f_* : \mathrm{Ext}(X) \to \mathrm{Ext}(Y)$, and $(fg)_* = f_* g_*$, $1_* = 1$.

(3) Suppose $A \subseteq X$ is closed, $\iota : A \to X$ is the inclusion map, $p : X \to X/A$ the quotient map. Let $\sigma \in \mathrm{Ext}(X)$ with $p_*\sigma = 0$. Then $\sigma = \iota_*\tau$ for some $\tau \in \mathrm{Ext}(A)$.

(4) Suppose X is the projective limit of the system (X_n, ρ_n) where ρ_n is onto for each n. Suppose $\sigma_n \in \mathrm{Ext}(X_n)$ for each n, and $(\rho_n)_*\sigma_{n+1} = \sigma_n$. Then there exists $\sigma \in \mathrm{Ext}(X)$ with $(\pi_n)_*\sigma = \sigma_n$.

(5) If $X \subseteq \underset{\sim}{R}$ then $\mathrm{Ext}(X) = 0$.

(6) There is a homomorphism $i : \mathrm{Ext}(X) \to G(X)$ which is natural in the sense that if $f : X \to Y$ is continuous, then $f_+ i(\sigma) = i(f_*\sigma)$, $\sigma \in \mathrm{Ext}(X)$.

(7) $i : \mathrm{Ext}(\Gamma) \to G(\Gamma) \approx \mathbb{Z}$ is an isomorphism (Γ = unit circle).

These properties hold whenever the spaces involved are compact metric - proofs can be found in [4]. We shall prove them for compact subsets of $\underset{\sim}{C}$ which simplifies matters and suffices for section 5.

4. Proof of Properties (1) - (7) for subsets of $\underset{\sim}{C}$

Property (1) is much the most difficult. We already know that $\mathrm{Ext}(X)$ is an Abelian semigroup so we have to show that it has an identity and that each element has an inverse.

Existence of an identity

We use formulation (A) of the definition. We start with a Lemma.

Lemma. Let T be essentially normal and let $\{\mu_k\}$ be a sequence of points of $\sigma_e(T)$. Then we can find an orthonormal sequence $\{f_k\}$ in H such that $\|(T-\mu_k)f_k\| \to 0$, $\|(T-\mu_k)^*f_k\| \to 0$.

Proof. We construct f_k inductively, so that for each k either $\|(T-\mu_k)f_k\| < 2^{-k}$ or $(T-\mu_k)^*f_k = 0$. Suppose $f_1 \dots f_k$ constructed. Since $\mu_{k+1} \in \sigma_e(T)$, either $T-\mu_k$ has infinite-dimensional kernel, or it fails to have closed range, or $(T-\mu_k)^*$ has infinite-dimensional kernel. In either of the first two cases we can choose $f_{k+1} \perp \mathrm{span}(f_1,\dots,f_k)$ with $\|f_{k+1}\| = 1$, $\|(T-\mu_{k+1})f_{k+1}\| < 2^{-k}$, in the third we can choose f_{k+1} with $(T-\mu_{k+1})^*f_{k+1} = 0$.

Now we have

$$\begin{aligned}&\|(T-\mu_k)f_k\|^2 - \|(T-\mu_k)^*f_k\|^2\\ &= \langle[(T-\mu_k)^*(T-\mu_k) - (T-\mu_k)(T-\mu_k)^*]f_k,f_k\rangle\\ &= \langle(T^*T-TT^*)f_k,f_k\rangle \to 0 \quad \text{as} \quad k \to \infty\end{aligned}$$

since $T^*T - TT^*$ is compact, and the desired result follows.

Now fix a compact subset X of $\underset{\sim}{C}$ and let D be a diagonal operator with eigenvalues $\lambda_1,\lambda_2 \dots \in X$ where each point of X is a limit point of the sequence $\{\lambda_n\}$ (i.e. there is an orthonormal basis $\{e_n\}$ of H such that $De_n = \lambda_n e_n$). Then $\sigma_e(D) = X$.

Proposition 4.1 If T is essentially normal and $\sigma_e(T) = X$, then $T \oplus D \sim T$.

Proof. Let $\{\mu_k\}$ be a sequence containing each of the points λ_n infinitely often. Let $\{f_k\}$ be as given by the Lemma. Let H_1 be the subspace spanned by $\{f_k\}$. Then with respect to the decomposition $H = H_1 \oplus H_1^{\perp}$ we have $T = (R \oplus S) + K$, where R is the diagonal operator on H_1 given by $Rf_k = \mu_k f_k$, $S \in B(H_1^{\perp})$, and $K \in B(H)$ is compact. So $T \sim R \oplus S$. Moreover $D \oplus R$ is unitarily equivalent to R (they are diagonal operators with the same eigenvalues) so $D \oplus T \sim D \oplus R \oplus S \sim R \oplus S \sim T$ as required.

It follows that $\mathrm{Ext}(X)$ has an identity element, namely $[D]$, which we denote by 0.

Before proceeding further, we show that if T is normal and $\sigma_e(T) = X$, then $[T] = 0$ - this is equivalent to Berg's theorem, that normal operators are classified completely up to $\sim$ by their essential spectra. First suppose T is diagonal, with eigenvalues $\lambda_1, \lambda_2, \ldots$. Then X is the set of limit points of $\{\lambda_n\}$. For each n choose $\mu_n \in X$ with $|\lambda_n - \mu_n| = \mathrm{dist}(\lambda_n, X)$. Then $|\lambda_n - \mu_n| \to 0$, so if D is the diagonal operator with eigenvalues $\mu_1, \mu_2, \ldots$, and the same eigenvectors as T, then $T - D$ is compact. Also by Prop. 4.1, $[D] = 0$, so $[T] = 0$.

Now the general case follows from:

Proposition 4.2 (Berg) Let T be normal and let $\epsilon > 0$. Then T can be expressed as $T = D + K$ where D is diagonal, K compact, and $\|K\| < \epsilon$.

Proof: First suppose T is multiplication by z on $L^2(\mu)$ where μ is a positive Borel measure on a bounded subset E of $\mathbb{C}$. For $k = 1,2,\ldots$, let $E_1^k, \ldots, E_{n_k}^k$ be a partition of E into disjoint Borel sets of diameter $< 2^{-k}\epsilon$, so that the $(k+1)$'th partition refines the k'th. Let M_k be the space of functions in $L^2(\mu)$ which are constant on E_j^k for $j = 1, \ldots, n_k$. Then $\{M_k\}$ is an increasing sequence of finite-dimensional subspaces whose union is dense in $L^2(\mu)$. Choose $\lambda_j^k \in E_j^k$ for each k & j, and define f_k on E by $f_k(z) = \lambda_j^k$ for $z \in E_j^k$. Let S_k be the operator on $L^2(\mu)$ of multiplication by f_k. Then M_r is a reducing subspace for S_k if $r \geq k$. Let

$$D = (S_1 | M_2) \oplus (S_2 | M_3 \ominus M_2) \oplus \ldots .$$

Now the norm of $(T-S_k)$ restricted to $M_{k+1} \ominus M_k$ is less than $2^{-k}\epsilon$, hence $T-D$ is compact and has norm $< \epsilon$. Also D is diagonal since each term in the direct sum is.

In the general case we can write $T = \oplus_n T_n$ where $T_1, T_2, \ldots$ are unitarily equivalent to multiplication operators of the type considered above. Then we can write $T_n = D_n \oplus K_n$ where D_n is diagonal, K_n is compact and $\|K_n\| < 2^{-n}\epsilon$. Then $T = D + K$ where $D = \oplus_n D_n$ is diagonal and $K = \oplus_n K_n$ is compact.

To sum up, $\mathrm{Ext}(X)$ has an identity element, which (as an equivalence class) includes all normal operators with essential spectrum X.

Existence of inverses

Here we use formulation (C) of the definition of $\mathrm{Ext}(X)$. In this formulation, the identity element of $\mathrm{Ext}(X)$ is represented by any $(1-1)$ *-homomorphism $\lambda : C(X) \to \mathcal{Q}$ of the form $\lambda = \pi \circ \rho$ where $\rho : C(X) \to B(H)$ is a *-homomorphism - because then $\lambda(\varphi) = \pi(\rho(\varphi))$ and $\rho(\varphi)$ is normal (φ being the function $\varphi(z) = z$).

We essentially follow an argument of Arveson, based on a C*-algebra lifting theorem of T.B. Andersen [1] and a dilation theorem of Neumark. As Andersens's theorem is complicated, we prove a weaker result which suffices for our purpose (Prop. 4.3, 4.4) and for the sake of completeness we sketch the proof of Neumark's theorem in the form we need.

Proposition 4.3 (Milutin) Let X be compact metric. Then there is a totally disconnected compact metric E, a *-homomorphism $P : C(X) \to C(E)$, and a positive linear map $Q : C(E) \to C(X)$ such that $QP = 1_{C(X)}$. (Positive means $f \geq 0 \Rightarrow Qf \geq 0$.)

Proof. Let F be the product of countably many copies of $\{-1, 1\}$, with the product topology. Let G be the product of countably many copies of Γ; G is a compact group. Let m be the product of normalized Lebesgue measure on the copies of Γ (= Haar measure on G). One can easily find a continuous map $\sigma : F \to G$ and a map $\tau : G \to F$ which is continuous almost everywhere (m), such that $\sigma\tau = 1_G$. (Do it first for Γ instead of G, e.g. by binary decimals, then take a countable product.) Denote the group operation in G by $+$.

Let $E = \{(x,y) \in F \times F : \sigma x + \sigma y \in X\}$. Assume $X \subseteq G$.

Define $P : C(X) \to C(E)$ by $Pf(x,y) = f(\sigma x + \sigma y)$, $f \in C(X)$.

We can write $(F \times F) \setminus E = \bigcup_{n=1}^{\infty} K_n$ where the K_n are closed, disjoint, and $\operatorname{diam}(K_n) \to 0$. For each n , choose $x_n \in E$, with $\operatorname{dist}(x_n, K_n) = \operatorname{dist}(E, K_n)$. For $g \in C(E)$ define $Rg \in C(F \times F)$ by

$$Rg(x) = \begin{cases} g(x) , & x \in E \\ g(x_n) , & x \in K_n \end{cases}$$

(the continuity of Rg is easily checked).

Finally define $Qg(\alpha) = \int_G Rg(\tau\beta, \tau(\alpha-\beta))dm(\beta)$, $\alpha \in X$.

That Qg is continuous on X follows from bounded convergence, because if $\alpha_n \to \alpha$, then $Rg(\tau\beta, \tau(\alpha_n-\beta)) \to Rg(\tau\beta, \tau(\alpha-\beta))$ for all β such that τ is continuous at $\alpha - \beta$, so $Qg(\alpha_n) \to Qg(\alpha)$. It is easily checked that P, Q have the required properties.

<u>Proposition 4.4</u> Let E be a totally disconnected compact metric space, A a C*-algebra, $I \subseteq A$ a closed ideal, and $\sigma : C(E) \to A/I$ a positive linear map with $\sigma 1 = 1$. Then there exists a positive linear map $\lambda : C(E) \to A$ with $\lambda 1 = 1$ and $\pi \circ \lambda = \sigma$.

We first prove a Lemma. A^+ denotes the set of positive elements of A .

<u>Lemma</u>. Let $T \in A^+$, let $f \,\&\, g \in (A/I)^+$ with $\pi(T) = f + g$, and let $\epsilon > 0$. Then there exist $F \,\&\, G \in A^+$ with $\pi(F) = f$, $\pi(G) = g$, and $\|F + G - T\| < \epsilon$.

<u>Proof</u>. Let $h = (f + g + \epsilon)^{-\frac{1}{2}} f (f + g + \epsilon)^{-\frac{1}{2}} \in A/I$.
Then $0 \leq h \leq 1$. Let R be a self-adjoint element of A with $\pi(R) = h$, and let $H = \varphi(R)$ where φ is defined by

$$\varphi(t) = \begin{cases} 0 , & t \leq 0 \\ t , & 0 \leq t \leq 1 \\ 1 , & t \geq 1 \end{cases}$$

Then $H \in A$, $0 \leq H \leq 1$, $\pi(H) = h$.
Now let $F = (T + \epsilon)^{\frac{1}{2}} H (T + \epsilon)^{\frac{1}{2}}$, $G_0 = (T + \epsilon)^{\frac{1}{2}} (I - H)(T + \epsilon)^{\frac{1}{2}}$. Then $F, G_0 \in A^+$, $\pi(F) = f$, $\pi(G_0) = g + \epsilon$, $F + G_0 = T + \epsilon$. Finally let $G = \psi(G_0)$ where $\psi(t) = \max(0, t-\epsilon)$, $t \in \mathbb{R}$. Then $G \in A^+$, $\pi(G) = g$,

and $\|G-G_o\| \leq \varepsilon$, so $\|F+G-T\| \leq 2\varepsilon$.

To complete the proof of Prop. 4.4, we consider for each $n = 0,1,2,\ldots$, a partition $\mathcal{P}_n$ of E into 2^n disjoint closed subsets, so that each set of $\mathcal{P}_n$ is the union of 2 sets of $\mathcal{P}_{n+1}$, and so that the diameter of the biggest set in $\mathcal{P}_n$ tends to 0 as $n \to \infty$.
Let Ω_n denote the $\leq 2^n$- dimensional subspace of $C(E)$ of functions that are constant on each set of $\mathcal{P}_n$. Let $\Omega = \bigcup_n \Omega_n$. Using the lemma, one can easily construct inductively positive linear maps $\mu_n : \Omega_n \to A$ such that $\pi \circ \mu_n = \sigma|\Omega_n$, $\|\mu_{n+1}(f) - \mu_n(f)\| < 2^{-n-2}\|f\|$, $f \in \Omega_n$, and $\mu_o 1 = 1$. Then for $f \in \Omega$, $\mu_n(f)$ converges to a limit $\mu(f)$, and $\pi \circ \mu = \sigma|\Omega$. μ is positive linear and $\|\mu(1) - 1\| \leq \frac{1}{2}$. Then $\pi(\mu(1)^{-\frac{1}{2}}) = 1$, so we can define

$$\lambda(f) = \mu(1)^{-\frac{1}{2}}\mu(f)\mu(1)^{-\frac{1}{2}}$$

then $\lambda : \Omega \to A$ is positive, $\lambda(1) = 1$, $\pi \circ \lambda = \sigma$, so λ is bounded and extends to $C(E)$ (Ω is dense in $C(E)$), with the same properties.

<u>Proposition 4.5</u> Let $\lambda : C(E) \to B(H)$ be positive linear with $\lambda 1 = 1$. Then there is a $*$-homomorphism $\tau : C(E) \to B(H \oplus H)$ such that the matrix of $\tau(f)$ has $\lambda(f)$ as its upper left corner.

<u>Proof</u>. Let K_o be the vector space of all formal expressions $\sum_{i=1}^{n} h_i F_i$ where $h_i \in H$ and F_i are clopen subsets of E , it being understood that $h_1 F + h_2 F = (h_1 + h_2)F$.
Equip K_o with a semi-inner product by

$$\langle \Sigma\, h_i F_i,\ \Sigma\, k_j G_j \rangle = \sum_{i,j} \langle \lambda(\chi_{F_i \cap G_j}) h_i, k_j \rangle .$$

We have to verify that $\sum_{i,j} \langle \lambda(\chi_{F_i \cap F_j}) h_i, h_j \rangle \geq 0$.
To do this choose disjoint clopen sets $E_1, \ldots, E_N$ so that $\chi_{F_i} = \sum_{k=1}^{N} \rho_{ik} \chi_{E_k}$ where $\rho_{ik} = 0$ or 1 . Let $T_k = \lambda(\chi_{E_k}) \geq 0$. Then

$$\begin{aligned} \sum_{i,j} \langle \lambda(\chi_{F_i \cap F_j}) h_i, h_j \rangle &= \sum_{i,j,k} \rho_{ik}\rho_{jk} \langle T_k h_i, h_j \rangle \\ &= \sum_k \langle T_k g_k, g_k \rangle \quad \text{where } g_k = \sum_i \rho_{ik} h_i \\ &\geq 0 . \end{aligned}$$

We get a Hilbert space K from K_0 by taking the quotient by the null space and completing in the usual way. We embed H isometrically in K by $h \to hE$.

For a clopen subset $F \subseteq E$, define an operator $\rho(F)$ on K by $\rho(F)(hG) = h(F \cap G)$. It is easily checked that $\rho(F)$ is a self-adjoint projection, and $\rho(F)\rho(G) = \rho(F \cap G)$. It follows that we can define a $*$-homomorphism $\tau : C(E) \to B(K)$ by $\tau(\chi_F) = \rho(F)$ and extending by linearity and continuity. For $h,k \in H$ we have $\langle \rho(F)h,k \rangle = \langle \lambda(\chi_F)h,k \rangle$, hence $\langle \tau(f)h,k \rangle = \langle \lambda(f)h,k \rangle$ for $f \in C(E)$, so $\lambda(f)$ is the compression of $\tau(f)$ to H. The desired result follows by identifying $K \ominus H$ with H. (If it happens that $K \ominus H$ is finite-dimensional, one can just add on another copy of H.)

(For more details of this construction, see [8].
Our situation is simpler because E is totally disconnected.)

We now apply these 3 propositions to prove that $\mathrm{Ext}(X)$ is a group when $X \subseteq \underset{\sim}{C}$. Let $\mu : C(X) \to \mathcal{Q}$ be a $(1-1)$ $*$-homomorphism. Let E, P, Q be as given by Prop. 4.3, and let $\sigma = \mu \circ Q : C(E) \to \mathcal{Q}$. Then Prop. 4.4 applies to σ and we get $\lambda : C(E) \to B(H)$, positive linear, $\lambda 1 = 1$, $\pi \circ \lambda = \sigma$. Now let $\tau : C(E) \to B(H \oplus H)$ be as given by Prop. 4.5. Let $\nu = \tau \circ P : C(X) \to B(H \oplus H)$ – then ν is a $*$-homomorphism. Let $\theta = \pi \circ \nu : C(X) \to \mathcal{Q}(H \oplus H)$. Then θ is a $*$-homomorphism, and the upper left entry in the matrix of $\theta(f)$ is $\pi \circ \lambda \circ P(f) = \mu(f)$, i.e. we can write $\theta(f) = \begin{pmatrix} \mu(f) & a(f) \\ b(f) & c(f) \end{pmatrix}$ say.

Now $\theta(\bar{f}f) = \theta(f)^*\theta(f)$ yields $b(f)^*b(f) = 0$, so $b(f) = 0$ and similarly $a(f) = 0$. It then follows that $f \to c(f)$ must be a $*$-homomorphism, and $\theta(f) = \mu(f) \oplus c(f)$, so $[\theta] = [\mu] + [c]$. But $[\theta]$ is 0 since $\theta = \pi \circ \nu$, so $[\mu]$ is invertible in $\mathrm{Ext}(X)$, as required.

This completes the proof of Property (1).

<u>Proof of Property (2)</u>

We use formulation (B). Let X and Y be compact subsets of $\underset{\sim}{C}$, and let $f : X \to Y$ be continuous. We define $f_* : \mathrm{Ext}(X) \to \mathrm{Ext}(Y)$ as follows: if x is a normal element of $\mathcal{Q}$ with spectrum X, then $f(x)$ is a normal element with spectrum $f(X) \subseteq Y$. Then if D is diagonal, with essential spectrum Y, $f(x) \oplus \pi(D)$ has spectrum Y. We define $f_*[x] = [f(x) \oplus \pi(D)]$. Using Prop.4.1 it is a routine exercise that f_* is well-defined, a group homomorphism, and that $(fg)_* = f_* g_*$ (when defined) and $1_* = 1$. We note also that if f is onto, we have $f_*[x] = [f(x)]$.

Proof of Property (3)

We prove (3) in the case when $X \subseteq \underset{\sim}{C}$ and X/A is homeomorphic to a subset of $\underset{\sim}{C}$. In this case (3) can be restated as follows:

Proposition 4.6 Let X and Y be compact subsets of $\underset{\sim}{C}$ and A a non-empty closed subset of X. Let $p : X \to Y$ be continuous and suppose p maps A to a single point y_0 and $X \backslash A$ $(1-1)$ onto $Y \backslash \{y_0\}$. Let $\iota : A \to X$ be the inclusion map.

Suppose $\sigma \in \text{Ext}(X)$, $p_* \sigma = 0$. Then $\sigma = \iota_* \tau$ for some $\tau \in \text{Ext}(A)$.

Proof. Use formulation (A). Let $\sigma = [T]$ where T is essentially normal, $\sigma_e(T) = X$. Then $p(\pi(T)) = \pi(D)$ where D is a diagonal operator with eigenvalues in Y. Let M_0 be the span of the eigenvectors of D whose eigenvalues satisfy $|\lambda - y_0| \geq 1$, and for $k = 1,2,\ldots$ let M_k be the span of those with $2^{-k} \leq |\lambda - y_0| < 2^{-k+1}$. Let P_k be the projection on M_k. So the P_k are mutually orthogonal.

Claim: For each k there is an operator S_k, of the form $f(D)$ where f is a continuous function on Y, such that $(T - S_k)P_k$ and $P_k(T - S_k)$ are compact. To prove this, fix k and choose continuous functions f and g on Y and h on X such that $z = f(p(z)) + h(z)g(p(z))$, $z \in X$, and such that $g(z) = 0$ whenever $|z - y_0| \geq 2^{-k}$. Then $\pi(T) = f(\pi(D)) + h(\pi(T))g(\pi(D))$. Now $g(\pi(D))\pi(P_k) = 0$ so $\pi(T)\pi(P_k) = f(\pi(D))\pi(P_k) = \pi(f(D)P_k)$ so with $S_k = f(D)$ we have $(T - S_k)P_k$ is compact and similarly $P_k(T - S_k)$ is compact, proving the claim.

Again with a fixed k, let $e_1, e_2, \ldots$ (possibly terminating) be the eigenvectors of D in M_k. Let V_j be the projection on the span of $\{e_i : i > j\}$. Then $\|V_j(T - S_k)\| \to 0$, $\|(T - S_k)V_j\| \to 0$ as $j \to \infty$. Choose j so that these norms are both $< 2^{-k}$, and define $Q_k = V_j$. Then $P_k - Q_k$ is a finite rank projection, Q_k commutes with D (and hence S_k), and $\|Q_k(T-S_k)\| < 2^{-k}$, $\|(T-S_k)Q_k\| < 2^{-k}$. Moreover, for different k the Q_k are mutually orthogonal.

Let $Q = \sum_{k=0}^{\infty} Q_k$, $S = \sum_{k=0}^{\infty} Q_k S_k$. Then Q is a projection, and $(T-S)Q = \Sigma(T-S_k)Q_k$ is compact, likewise $Q(T-S)$. Thus $T - [S + (1-Q)T(1-Q)]$ is compact, so with respect to the decomposition $H = QH \oplus (1-Q)H$ we have $T = S \oplus R + K$ where K is compact and R is some operator on $(1-Q)H$. Since T is essentially normal, so is R.

Finally, $\sigma_e(R) \subseteq A$. To see this, let W_k be the restriction of S to the range of $\sum_{j>k} Q_j$. Then $p(\pi(W_k \oplus R))$ is π (restric- of D to span of eigenvectors with eigenvalues in $|\lambda - y_0| \le 2^{-k}$) so has spectrum in $|\lambda - y_0| \le 2^{-k}$. So $p(\sigma_e(R)) \subseteq \{z : |z-y_0| \le 2^{-k}\}$ for all k, so $\sigma_e(R) \subseteq A$.

So we have $T \sim S \oplus R$ where S is diagonal, with $\sigma_e(S) \subseteq X$, and R is essentially normal with $\sigma_e(R) \subseteq A$. If now D_1 and D_2 are diagonal with essential spectra X & A respectively, then by Prop. 4.1, $T \sim (S \oplus D_1) \oplus (R \oplus D_2)$ so $\sigma = [T] = \iota_*[R \oplus D_2]$ as required.

Proof of Property (4) (valid for any compact metric spaces).

Use formulation (C). Let $\lambda_n : C(X_n) \to \mathcal{Q}$ be a (1 - 1) *-homomorphism representing σ_n. We may assume $\lambda_n(f) = \lambda_{n+1}(f \circ \rho_n)$, $f \in C(X_n)$ (if necessary adjusting $\lambda_2, \lambda_3, \ldots$ successively by unitary operators). Let Ω_n be the subspace of functions in $C(X)$ of the form $f \circ \pi_n$, $f \in C(X_n)$. Then $\{\Omega_n\}$ is an increasing sequence of subalgebras of $C(X)$ whose union Ω is dense in $C(X)$. Define $\lambda(f \circ \pi_n) = \lambda_n(f)$. This definition is consistent and defines a *-homomorphism $\lambda : \Omega \to \mathcal{Q}$ which extends by continuity to $C(X)$. Moreover λ is (1 - 1) because if the ideal $\ker \lambda$ were non-trivial, it would have a non-trivial intersection with Ω_n for some n, which would contradict λ_n being (1 - 1). Then $\sigma = [\lambda]$ is the required element of $\mathrm{Ext}(X)$.

Proof of Property (5)

Let T be essentially normal, with $\sigma_e(T) = X \subseteq \mathbb{R}$. Then $\pi(T)$, being normal with real spectrum in $\mathcal{Q}$, is self-adjoint. So

$\pi(T-T^*) = 0$, so $T - T^*$ is compact, so $T \sim \frac{1}{2}(T+T^*)$. But $\frac{1}{2}(T+T^*)$ is self-adjoint, so $[T] = 0$ by Prop. 4.2.

Proof of Property (6)

For $\sigma \in \mathrm{Ext}(X)$, let $\lambda : C(X) \to \mathcal{A}$ represent σ . For $f \in C(X)^{-1}$, define $i(\sigma)(f) = \mathrm{ind}\, \lambda(f)$. This depends only on the coset of f in $B(X)$ as described at the end of section 1, so $i(\sigma)$ gives a homomorphism from $B(X)$ to $\mathbb{Z}$, i.e. an element of $G(X)$. That i is a homomorphism, and the naturality property, are routine verifications.

Proof of Property (7)

Suppose T is essentially normal, $\sigma_e(T) = \Gamma$, and $\mathrm{ind}(T) = 0$. Then $\pi(\tilde{T})$, being normal with spectrum Γ , is unitary. So $1 - T^*T$ and $1 - TT^*$ are compact. Let $T = VP$ be the polar decomposition of T , where V is a partial isometry and $P = (T^*T)^{\frac{1}{2}}$. Then $\pi(P) = (\pi(T^*T))^{\frac{1}{2}} = 1$, so $T - V$ is compact, so $T \sim V$. T , and hence also V , is Fredholm with index 0, so the domain and range of V have the same finite codimension. Hence by adding to V a finite-rank partial isometry, we can get a unitary operator U . By Prop. 4.2, $[U] = 0$, so $[T] = [V] = [U] = 0$.

This shows that $i : \mathrm{Ext}(\Gamma) \to G(\Gamma)$ is (1 - 1). Since the index of the unilateral shift is -1 , i is also onto.

5. Proof of the Theorem

In this section we prove the theorem stated in Section 1, which is equivalent to the assertion that for $X \subseteq \mathbb{C}$, $i : \mathrm{Ext}(X) \to G(X)$ is (1 - 1). The proof will be based solely on properties (1) to (7) - one can forget the definition of $\mathrm{Ext}(X)$.

The idea of the proof is this: given an element $\sigma \in \mathrm{Ext}(X)$ with $i(\sigma) = 0$, one cuts X into two pieces by a straight line, and shows that σ can be expressed as the sum of two elements, one coming from each piece of X (lemma 4). This process is repeated, cutting up X into smaller and smaller pieces. By a projective limit argument one shows that σ is then the image of an element of Ext of a totally disconnected space, which must be 0 .

We start with a couple of technical lemmas.

<u>Lemma 5.1</u>. Property (4) is valid without the assumption that ρ_n is onto.

<u>Proof</u>. For each n let Y_n be the union of X_n and a totally disconnected compact set F_n, disjoint from X_n. Extend the given map ρ_n on X_{n+1} to Y_{n+1}, so that it maps Y_{n+1} onto Y_n. Let Y be the projective limit of (Y_n, ρ_n). Then Y contains X and $Y \setminus X$ is totally disconnected (hence so is Y/X). Let $\iota : X \to Y$ and $\iota_n : X_n \to Y_n$ be the inclusion maps. By Property (4), there is $\tau \in \mathrm{Ext}(Y)$ with $(\pi_n)_* \tau = (\iota_n)_* \sigma_n$. Since Y/X is totally disconnected, and hence homeomorphic to a subset of $\underset{\sim}{R}$, $\mathrm{Ext}(Y/X) = 0$ by Property (5), so by Property (3) we can find $\sigma \in \mathrm{Ext}(X)$ with $\iota_* \sigma = \tau$. Then $(\iota_n)_*((\pi_n)_* \sigma - \sigma_n) = 0$. But $(\iota_n)_*$ is $(1-1)$, because there is a continuous map $r_n : Y_n \to X_n$ with $r_n \iota_n = 1$ so $(r_n)_* (\iota_n)_* = 1$. Thus $(\pi_n)_* \sigma = \sigma_n$ as required.

<u>Lemma 5.2</u>. Let $X = \bigcup_{n=1}^{\infty} J_n$, where J_n is a circle in $\underset{\sim}{C}$, $\mathrm{diam}(J_n) \to 0$, each J_n contains x_0, but $\{J_n \setminus \{x_0\}\}$ are mutually disjoint. Then $i : \mathrm{Ext}(X) \to G(X)$ is $(1-1)$.

<u>Proof</u>. Let $Y_n = \bigcup_{k>n} J_k$. Let $p : X \to X/Y_1$ be the quotient map. Let $\sigma \in \mathrm{Ext}(X)$ with $i(\sigma) = 0$. Then $i(p_* \sigma) = 0$, and since X/Y_1 is homeomorphic to J_1, it follows from Property (7) that $p_* \sigma = 0$. Hence by Property (3), $\sigma = (\iota_1)_* \sigma_1$, for some $\sigma_1 \in \mathrm{Ext}(Y_1)$, where $\iota_n : Y_n \to Y_{n-1}$ is the inclusion map. $(Y_0 = X)$. Now we have $i(\sigma_2)(\varphi_\alpha) = i(\sigma)(\varphi_\alpha) = 0$ for $\alpha \notin X$, hence by continuity for $\alpha \notin Y_1$ $(\varphi_\alpha(z) = z - \alpha)$. So $i(\sigma_1) = 0$. So we can repeat the process and get $\sigma_1 = \iota_2 \sigma_2$, $\sigma_2 \in \mathrm{Ext}(Y_2)$, ... Since the projective limit of the system (Y_n, ι_n) is a point, we conclude from Lemma 5.1 that $\sigma = 0$, as required.

<u>Lemma 5.3</u>. Let $X = X_1 \cup X_2$ where X_1 and X_2 are compact and $r : X \to X_1$ is a retraction of X onto X_1 with $r(X_2) \subseteq X_2$. Let $\iota_i : X_i \to X$ be the inclusion maps. Let $\sigma \in \mathrm{Ext}(X)$. Then $\sigma = (\iota_1)_* \sigma_1 + (\iota_2)_* \sigma_2$ for some $\sigma_i \in \mathrm{Ext}(X_i)$.

<u>Proof</u>. Let $\tau = \sigma - (\iota_1)_* r_* \sigma \in \mathrm{Ext}(X)$.

Let $p : X \to X/X_2$ be the quotient map. Then $p = p\iota_1 r$ so $p_*\tau = 0$. Hence by Property (3), $\tau = (\iota_2)_*\sigma_2$ for some $\sigma_2 \in \operatorname{Ext}(X_2)$. So with $\sigma_1 = r_*\sigma$ we have the required result.

Lemma 5.4. Let X be a compact subset of $\underset{\sim}{C}$, and let $\sigma \in \operatorname{Ext}(X)$ with $i(\sigma) = 0$. Let $a \in \underset{\sim}{R}$, let $X_1 = \{z \in X : \operatorname{Re} z \leq a\}$, $X_2 = \{z \in X : \operatorname{Re} z \geq a\}$. Let $\iota_i : X_i \to X$ be inclusion. Then we have $\sigma = (\iota_1)_*\sigma_1 + (\iota_2)_*\sigma_2$ where $\sigma_i \in \operatorname{Ext}(X_i)$, $i(\sigma_i) = 0$.

Proof. Choose a large number M and let $L = \{a + iy : -M \leq y \leq M\}$. Let $\iota : X \to X \cup L$, $\iota_3 : X_1 \cup L \to X \cup L$, $\iota_4 : X_2 \cup L \to X \cup L$, $\iota_5 : X_1 \to X_1 \cup L$, $\iota_6 : X_2 \to X_2 \cup L$ be the respective inclusions. Then Lemma 5.3, applied to $X \cup L = (X_1 \cup L) \cup (X_2 \cup L)$, yields $\mu_1 \in \operatorname{Ext}(X_1 \cup L)$, $\mu_2 \in \operatorname{Ext}(X_2 \cup L)$ with $\iota_*\sigma = (\iota_3)_*\mu_1 + (\iota_4)_*\mu_2$. Now if $\operatorname{Re} \alpha > a$, $\alpha \notin X$, then $i(\mu_1)(\varphi_\alpha) = 0$ and since $i(\sigma) = 0$ we have $i(\mu_2)(\varphi_\alpha) = 0$. Similarly for $\operatorname{Re} \alpha < a$, so we conclude $i(\mu_1) = 0$, $i(\mu_2) = 0$.

Now let $q : X_1 \cup L \to (X_1 \cup L)/X_1$ be the quotient map. Then $i(q_*\mu_1) = 0$, and since $X_1 \cup L/X_1 = L/X_1 \cap L$ is homeomorphic to a set of the type considered in Lemma 5.2, it follows that $q_*\mu_1 = 0$. So $\mu_1 = (\iota_5)_*\sigma_1$ for some $\sigma_1 \in \operatorname{Ext}(X_1)$. Similarly $\mu_2 = (\iota_6)_*\sigma_2$.

Then $\iota_*\sigma = \iota_*(\iota_1)_*\sigma_1 + \iota_*(\iota_2)_*\sigma_2$. Let $\rho = \sigma - (\iota_1)_*\sigma_1 - (\iota_2)_*\sigma_2 \in \operatorname{Ext}(X)$. Then $\iota_*\rho = 0$. So if $p : X \to X/X \cap L$ is the quotient map $p_*\rho = 0$. Since $\operatorname{Ext}(X \cap L) = 0$, it follows from Property (3) that $\rho = 0$. So $\sigma = (\iota_1)_*\sigma_1 + (\iota_2)_*\sigma_2$. Also $i(\sigma_i) = 0$ by the same argument as applied to μ_i.

Theorem Let $X \subseteq \underset{\sim}{C}$. Then $i : \operatorname{Ext}(X) \to G(X)$ is (1 - 1).

Proof. Let $\sigma \in \operatorname{Ext}(X)$, with $i(\sigma) = 0$. Let Q_0 be a closed square containing X. For $n = 1,2,\ldots$ let Q_n be a disjoint union of 4^n closed squares, $\rho_n : Q_{n+1} \to Q_n$ is onto, mapping each square of Q_{n+1} onto a quarter of a square in Q_n. Let $Y_n = \rho_{n-1}^{-1} \cdots \rho_0^{-1}(X)$. Let Y be the projective limit of the system (Y_n, ρ_n).

Using Lemma 5.4 twice (once for a vertical line and once for a horizontal line) we can find $\sigma_1 \in \operatorname{Ext}(Y_1)$ with $i(\sigma_1) = 0$ and $\rho_0(\sigma_1) = \sigma$. Repeating the process, we find $\sigma_n \in \operatorname{Ext}(Y_n)$ with

$\rho_n \sigma_{n+1} = \sigma_n$. Then by Property (4), there is $\tau \in \text{Ext}(Y)$ with $(\pi_0)_* \tau = \sigma$. But Y is totally disconnected, so $\text{Ext}(Y) = 0$ by Property (5), so $\sigma = 0$. Thus i is (1-1).

In fact i is also onto. Here is the proof, modulo some topology. Let $\mu \in G(X)$. Let $\{X_n\}$ be a decreasing sequence of "nice" compact sets (say finite unions of squares), with intersection X. Let $\iota_n : X_n \to X$ be inclusion, $\mu_n = (\iota_n)_+ \mu$. It is a topological fact that $G(X_n)$ is generated by elements of the form $f_+ \nu$ for $\nu \in G(\Gamma)$, $f : \Gamma \to X_n$ continuous. It follows that $\mu_n = i(\sigma_n)$ for some $\sigma_n \in \text{Ext}(X_n)$. If $\rho_n : X_{n+1} \to X_n$ is inclusion, the fact that i is (1-1) implies $(\rho_n)_* \sigma_{n+1} = \sigma_n$. So by Lemma 5.1 there is $\sigma \in \text{Ext}(X)$, $(\iota_n)_* \sigma = \sigma_n$, and then $(\iota_n)_+ (i(\sigma)) = \mu_n$ for each n which implies $i(\sigma) = \mu$.

This completes the classification of essentially normal operators up to $\sim$ – to specify an equivalence class one specifies a compact subset X of $\mathbb{C}$ and an integer for each bounded component of $\mathbb{C} \setminus X$.

6. Ext(X) for general X

Ext(X) can be defined for arbitrary compact metric X using formulation (C). Here we shall describe some aspects of this general theory without any attempt at a systematic discussion.

As mentioned before, Properties (1) to (7) are valid in this setting. The only part whose proof requires significant alteration is the existence of an identity for Ext(X), which is represented by any (1-1) *-homomorphism $\lambda : C(X) \to \mathcal{A}$ of the form $\lambda = \pi\mu$ where $\mu : C(X) \to B(H)$ is a *-homomorphism. This is treated in [4, Section 5]. Given this, the proofs of (1) to (7) go through without essential change, except that (3) requires some technical modifications. We shall henceforth assume (1) to (7).

When X is a subset of $\mathbb{R}^n$, Ext(X) can be interpreted in terms of commuting n-tuples of self-adjoint elements of $\mathcal{A}$. If $\alpha_1, \ldots, \alpha_n$ are commuting self-adjoint elements of $\mathcal{A}$, their joint spectrum X (in the commutative C*-algebra they generate) is a subset of $\mathbb{R}^n$, and $f \to f(\alpha_1, \ldots, \alpha_n)$ defines a (1-1) *-homomorphism from $C(X) \to \mathcal{A}$. Conversely, given a (1-1) *-homomorphism $\lambda : C(X) \to \mathcal{A}$, let φ_i be the i'th coordinate function on X, then $\alpha_i = \lambda(\varphi_i)$ are commuting self-adjoint elements of $\mathcal{Q}$, with joint spectrum X, and $\lambda(f) = f(\alpha_1,\ldots,\alpha_n)$. $(\alpha_1,\ldots,\alpha_n)$ and $(\beta_1,\ldots,\beta_n)$ give the same element of $\operatorname{Ext}(X)$ if and only if there is a unitary U with $\alpha_i = \pi(U)^{-1}\beta_i\pi(U)$, $i = 1,\ldots,n$.

The problem of normal elements of $\mathcal{Q}$ is just the case $n = 2$ of the above, since a normal element can be identified with a pair of commuting self-adjoint elements. If $X \subseteq \underset{\sim}{C}^n$, we can interpret $\operatorname{Ext}(X)$ in terms of commuting n-tuples of normal elements of $\mathcal{Q}$.

We shall describe 3 special cases which illustrate some of the features of the general theory, using only properties (1) to (7).

6.1 The unit sphere in $\underset{\sim}{C}^n$

This gives the simplest example where $i : \operatorname{Ext}(X) \to G(X)$ is not (1 - 1). We consider the case $n = 2$, and let $S = \{(z,w): |z|^2+|w|^2=1\}$ be the unit sphere in $\underset{\sim}{C}^2$. Let m be normalized surface measure on S, and let H^2 be the closure in $L^2(m)$ of the space of polynomials in z and w. For $f \in C(S)$ let $\lambda(f)$ be the Toeplitz operator on H^2 defined by $\lambda(f)g = P(fg)$, $g \in H^2$, where P is the projection of L^2 onto H^2. Then $\lambda(f)^* = \lambda(\bar{f})$. Also $\lambda(f_1f_2) - \lambda(f_1)\lambda(f_2)$ is compact (this can be shown as follows - it is enough to verify it when f_1 and f_2 are polynomials in $z,\bar{z},w,\bar{w}$, because they are dense in $C(S)$. This will follow easily if we can show $\lambda(z),\lambda(\bar{z}),\lambda(w)$, $\lambda(\bar{w})$ commute with each other modulo the compacts. This is a matter of routine checking using the orthonormal basis $e_{kl} = \left[\frac{(k+l+1)!}{k!l!}\right]^{\frac{1}{2}} z^k w^l$ $(k,l \geq 0)$ for H^2 and noting that $\lambda(z)e_{kl} = \left(\frac{k+1}{k+l+2}\right)^{\frac{1}{2}} e_{k+1,l}$, $\lambda(\bar{z})e_{kl} = \left(\frac{k}{k+l+1}\right)e_{k-1,l}$ (interpreting $e_{-1,l}$ as 0) with similar formulae for $\lambda(w)$, $\lambda(\bar{w})$).

Thus, if we put $\mu(f) = \pi\lambda(f)$, then $\mu : C(S) \to \mathcal{Q}$ is a *-homomorphism, which is easily shown to be (1 - 1), so $[\mu]$ is an element of $\operatorname{Ext}(S)$.

We claim that $[\mu] \neq 0$. To see this, consider the operator T on $H^2 \oplus H^2$ given by the matrix $\begin{bmatrix} \lambda(z) & \lambda(w) \\ -\lambda(\bar{w}) & \lambda(\bar{z}) \end{bmatrix}$. We see that $\pi(T^*T) = \pi(TT^*) = 1$ so $\pi(T)$ is unitary, so T is Fredholm. If $[\mu] = 0$ then T is a compact perturbation of an operator of the form

$N = \begin{bmatrix} N_1 & N_2 \\ -N_2^* & N_1^* \end{bmatrix}$ where N_1 & N_2 are commuting normal operators. But then N is normal so the index of T is 0. We show that in fact $\mathrm{ind}(T) < 0$. First, $\ker T = 0$, for if $(f,g) \in \ker T$ then (1) $zf + wg = 0$ and (2) $P(-\bar{w}f + \bar{z}g) = 0$. From (1), and looking at expansions of f & g w.r.t. (e_{kl}), we find $h \in H^2$ with $f = wh$, $g = -zh$. Then (2) gives $Ph = 0$, so $h \perp H^2$, so $h = 0$, whence $(f,g) = 0$. So $\ker T = 0$. On the other hand, $(1,0) \in \ker T^*$, so $\mathrm{ind}(T) < 0$.

Thus we have $[\mu] \neq 0$. On the other hand $G(S) = 0$ (topological fact) so i is not $(1-1)$ in this case.

It will be observed that we have still used the concept of index to show that $[\mu] \neq 0$. This suggests that a generalization of our homomorphism $i : \mathrm{Ext}(X) \to G(X)$ might be found under which $[\mu]$ would have non-trivial image. This can be done as follows: suppose ρ is a continuous map from X to the group of invertible $n \times n$ complex matrices - or equivalently an invertible $n \times n$ matrix (ρ_{ij}) whose entries are in $C(X)$. Then if $\lambda : C(X) \to \mathcal{Q}$ represents an element of $\mathrm{Ext}(X)$, then the matrix $\lambda(\rho_{ij})$ can be regarded as an invertible element of the Calkin algebra on the direct sum of n copies of H, and as such has an index which we denote $i_n([\lambda])(\rho)$. This gives a homomorphism $i_n : \mathrm{Ext}(X) \to G_n(X)$, where $G_n(X)$ is the group of homomorphisms from the group of invertible continuous $(n \times n)$-matrix valued functions on X, modulo its principle component, to $\mathbb{Z}$.

In the case of the sphere in $\underset{\sim}{C}^n$, $i_n : \mathrm{Ext}(X) \to G_n(X) \approx \mathbb{Z}$ can be shown to be an isomorphism. However, in general the i_n are not enough to reveal the full structure of $\mathrm{Ext}(X)$, because as the next example shows, $\mathrm{Ext}(X)$ may contain elements of finite order, which i_n necessarily maps to 0 (because $G_n(X)$ can have no elements of finite order).

6.2 The real projective plane

We show that if P is the real projective plane, then $\mathrm{Ext}(P)$ contains an element of order 2. Topologically, P may be obtained from a Møbius band M by identifying the edge of M to a point, so we consider first $\mathrm{Ext}(M)$.

There are two natural embeddings of the circle Γ into M, namely $\alpha : \Gamma \to M$ which maps Γ onto the centre line of M, and $\beta : \Gamma \to M$ which goes round the edge. Also the retraction of M onto its centre line gives $r : M \to \Gamma$ with $r\alpha = 1$ and $r\beta(e^{i\theta}) = e^{2i\theta}$ (because the edge goes round twice). Now $M/\alpha(\Gamma)$ is a disc, so $\mathrm{Ext}(M/\alpha(\Gamma)) = 0$, so α_* is onto by Property (3). Since $r_*\alpha_* = 1$ it follows that α_* is an isomorphism between $\mathrm{Ext}(\Gamma) = \mathbb{Z}$ and $\mathrm{Ext}(M)$.

If σ is a generator of $\mathrm{Ext}(\Gamma)$, we see that $\alpha_*\sigma$ is a generator of $\mathrm{Ext}(M)$ and $\beta_*\sigma = 2\alpha_*\sigma$ (since $r_*\beta_*\sigma = 2\sigma$). So the image of β_* is $\{2n\alpha_*\sigma : n \in \mathbb{Z}\}$.

Now $P = M/\beta(\Gamma)$. Let $q : M \to P$ be the quotient map. Now the kernel of q_* is just the image of β_*, so $q_*\alpha_*\sigma \neq 0$, but $q_*(2\alpha_*\sigma) = 0$. So $q_*\alpha_*\sigma$ is an element of $\mathrm{Ext}(P)$ of order 2.

In operator terms this can be expressed as follows: there is a pair (T_1, T_2) of essentially normal operators, commuting modulo the compact operators, such that the pair $(T_1 \oplus T_1, T_2 \oplus T_2)$ can be expressed in the form $(N_1 + K_1, N_2 + K_2)$ where N_1 and N_2 are commuting normal operators and K_1 and K_2 are compact, but the pair (T_1, T_2) cannot be so expressed.

We can describe the operators T_1 and T_2 explicitly: embed P in $\mathbb{C}^2$ in such a way that the circle $q\alpha(\Gamma)$ is the unit circle in the plane $\mathbb{C} \times \{0\}$. Let (D_1, D_2) be a pair of commuting diagonal operators with joint essential spectrum P. Let T denote the unilateral shift. Let $T_1 = T \oplus D_1$, $T_2 = 0 \oplus D_2$. Then (T_1, T_2) represents the element $q_*\alpha_*\sigma$ in $\mathrm{Ext}(P)$, so has the properties of the previous paragraph.

Finally we remark that it can be shown that $\mathrm{Ext}(P) \approx \mathbb{Z}_2$.

6.3 The suspended solenoid

Here we give an example to show that the analogue of the result "$N + K$ is closed" for pairs is false.

We first note that the real projective plane is homeomorphic to the closed unit disc, with z identified to $-z$ for $z \in \Gamma$. We may choose the homeomorphism so that the embedding $q\alpha$ of Γ into P obtained in the last section maps Γ onto $[-1, 1]$ (which, with -1 identified to 1, is a circle). We call this embedding ψ, so

$\psi_*\sigma \neq 0$ in $\mathrm{Ext}(P)$.

We now introduce the (triadic) solenoid $\tilde{\Gamma}$ as the set of all sequences $\underset{\sim}{z} = (z_1, z_2, \dots)$ with $z_n \in \Gamma$ and $z_n = z_{n+1}^3$, with the product topology. We write $-\underset{\sim}{z} = (-z_1, -z_2, \dots)$. We denote by $\tilde{P}$ the space obtained from $\mathrm{cone}(\tilde{\Gamma})$ by identifying $(1,\underset{\sim}{z})$ with $(1,-\underset{\sim}{z})$. We define $\alpha : \tilde{\Gamma} \to \tilde{P}$ by $\alpha(\underset{\sim}{z}) = (1,(z_1^{\frac{1}{2}}, z_2^{\frac{1}{2}}, \dots))$, and let $q : \tilde{P} \to \tilde{P}/\alpha(\tilde{\Gamma}) = S\tilde{\Gamma}$ be the quotient map. ($S\tilde{\Gamma}$ is the suspension of $\tilde{\Gamma}$). Define $\mu : \tilde{\Gamma} \to \Gamma$ by $\mu(\underset{\sim}{z}) = z_1$, $\nu : \tilde{P} \to P$ by $\nu(t,\underset{\sim}{z}) = tz_1$, and $\beta : \Gamma \to P$ by $\beta(z) = z^{\frac{1}{2}}$. Then $\beta\mu = \nu\alpha$.

We distinguish two points of $\tilde{\Gamma}$: $\gamma^+ = (1,1,1,\dots)$, $\gamma^- = (-1,-1,\dots)$. Then there is a map $\tilde{\psi} : \Gamma \to \tilde{P}$, whose range is the circle $\{(t,\gamma^+)\} \cup \{(t,\gamma^-)\}$, such that $\psi = \nu\tilde{\psi}$. Let $\tau = \tilde{\psi}_*\sigma$, where σ is a generator of $\mathrm{Ext}(\Gamma)$. Then $\nu_*\tau \neq 0$, so $\tau \neq 0$. Moreover, $\tau \notin \mathrm{Im}(\alpha_*)$ because it is easily seen that $\mu_* = 0$, so $\nu_*\alpha_* = \beta_*\mu_* = 0$. Hence by property (3), $\theta = q_*\nu_*\tau \neq 0$.

We embed $\tilde{\Gamma}$ in $\underset{\sim}{R}^3$ (for example, as the intersecting of a decreasing sequence of solid tori, each running round 3 times inside the preceeding one). We can then embed $S\tilde{\Gamma}$ in $\underset{\sim}{C}^2$. Since $q\nu$ maps Γ onto $S\{\gamma^+,\gamma^-\}$, we can arrange this embedding so that $q\nu$ maps Γ to $\Gamma \times \{0\}$ in $\underset{\sim}{C}^2$.

Let $\{X_n\}$ be a decreasing sequence of closed neighbourhoods of $S\tilde{\Gamma}$, with intersection $S\tilde{\Gamma}$. Let (z_k,w_k) be a dense sequence in $S\tilde{\Gamma}$, and (z_{nk},w_{nk}) a dense sequence in X_n , so that as $n \to \infty$, $\sup_k \max(|z_k - z_{nk}|, |w_k - w_{nk}|) \to 0$. With respect to a fixed orthonormal basis $\{e_k\}$, let D, C, D_n, C_n be respectively the diagonal operators with eigenvalues z_k, w_k, z_{nk}, w_{nk} .

The pair $(T \oplus D, 0 \oplus C)$, where T is the unilateral shift, represents $\theta \neq 0$ in $\mathrm{Ext}(S\tilde{\Gamma})$, and therefore is not of the form $(N_1 + K_1, N_2 + K_2)$ where N_1, N_2 are commuting normal operators and K_1, K_2 are compact. On the other hand, the embedding $\Gamma \times \{0\} \to X_n$ factors through the closed unit disc, and hence induces the zero homomorphism of Ext . So the pair $(T \oplus D_n, 0 \oplus C_n)$ is of the form $(N_1 + K_1, N_2 + K_2)$. Since $D_n \to D$ and $C_n \to C$, the set of pairs of the form $(N_1 + K_1, N_2 + K_2)$ is not norm closed.

6.4 The general theory of Ext(X)

The full theory developed by Brown, Douglas and Fillmore describes Ext(X) in terms of known algebraic topological objects. Basically, it is a homology theory dual to K-theory. We make no attempt to explain this statement, but merely mention some relevant facts:

(a) Properties (1) to (7) are not sufficient to describe Ext(X) completely, but from (1) to (5) it is possible to deduce homotopy invariance and an exact homology sequence for Ext .

(b) There is a natural isomorphism of Ext(X) onto $\mathrm{Ext}(S^2X)$ (S^2X is the second suspension of X , S(SX)), analogous to the Bott periodicity theorem of K-theory. In particular, $\mathrm{Ext}(S^n)$ is 0 if n is even, $\mathbb{Z}$ if n is odd. $\mathrm{Ext}(S^{2k-1})$ is generated by the element arising from the Toeplitz operators in $\underset{\sim}{C}^k$, as described in 6.1. The proofs of these facts are very difficult.

(c) $i : \mathrm{Ext}(X) \to G(X)$ is an isomorphism if $X \subseteq \underset{\sim}{R}^3$. This can be proved by methods similar to those of Section 5, once one knows that $\mathrm{Ext}(S^2) = 0$. As a consequence, the set of all triples $(T_1 + K_1, T_2 + K_2, T_3 + K_3)$, where T_1, T_2, T_3 are commuting self-adjoint operators and K_1, K_2, K_3 are compact self-adjoint operators, is norm closed. The example in 6.3 shows that this is false for 4-tuples.

References

1. T.-B. Andersen, Linear extensions, projections and split faces, J. Functional Anal. 17 (1974), 161-173.

2. I.D. Berg, An extension of the Weyl - von Neumann theorem to normal operators, Trans.Amer.Math.Soc. 160 (1971), 365-371.

3. L.G. Brown, R.G. Douglas and P.A. Fillmore, Extensions of C*-algebras, operators with compact self-commutators, and K-homology, Bull.Amer.Math.Soc. 79 (1973), 973-978.

4. L.G. Brown, R.G. Douglas and P.A. Fillmore, Unitary equivalence modulo the compact operators and extensions of C*-algebras, Springer Lecture Notes in Mathematics, No.345, 58-128.

5. J.A. Deddens and J.G. Stampfli, On a question of Douglas and Fillmore, Bull.Amer.Math.Soc. 79 (1973), 327-330.

6. R.G. Douglas, Banach Algebra Techniques in Operator Theory, Academic Press, New York, 1972.

7. N. Dunford and J.T. Schwartz, Linear Operators, Part II, Wiley Interscience

8. B. Sz.-Nagy, Extensions of linear transformations in Hilbert space which extend beyond this space, Appendix to F. Riesz and B. Sz.-Nagy, Functional Analysis (New York, 1960).

FINE POTENTIAL AND FUNCTION ALGEBRAS

A. Debiard, B. Gaveau

1. Introduction

Let A be a function algebra, X its Silov boundary, M its maximal ideal space. One of the outstanding problems in function theory is to find a structure of complex analytic space on M such that the functions of A are analytic for this analytic structure. In some cases (algebras with few representative measures, or polynomial hulls of regular curves in $\mathbb{C}^n$), such structures can be described. (See also the recent work of R. Basener [1] and of Gamelin, Sibony and Wermer, this conference.) The first counterexample to the existence of analytic structure was found by Stolzenberg [14] who exhibited a compact $K \subset \mathbb{C}^2$ such that the polynomial hull $h(K)$ of K is distinct from K, and there exists no analytic structure on $h(K) - K$. More explicit examples were found by Wermer and Basener for rationally convex hulls (see [2] and also [6] for related class of examples). Moreover, if an analytic structure exists, it would have to live on the Gleason parts of the maximal ideal space. As a consequence, if f is a non constant function in a Gleason part P, then $f(P)$ would have to be interior point to $f(M)$, by the fact that analytic functions are open mappings. And that assertion is completely false in general.

The interest for the research of analytic structure is that it provides regularity theorems on the functions of A, and also theorems about representative measures, bounded point derivation, approximation theory, etc. But such theorems, in particular differentiability theorems about the functions of A, can be proved if one can find in the maximal ideal space, some kind of dynamical system along the trajectories of which one can study the properties of functions. The aim of this paper is to show how such generalized dynamical systems (brownian paths or even sheets) lives in spectrum of certain function algebras even if there is no analytic structure in these spectrum, and to obtain from that regularity theorems, theorems about Jensen measures, approximation and derivations, just as if there was analytic structure.

2. Adapted diffusions, Jensen parts, Jensen boundary

Let us call $m_\omega(t)$ an adapted diffusion in the maximal ideal space M of A, if $m_\omega(t)$ is a strong Markov process with continuous paths such that for each $f \in A$, $f(m_\omega(t))$ is a complex conformal martingale, in the sense that it is a change of time of a complex brownian motion. For example, if M carries some analytic structure, then any Kähler process on this structure (see [11]) is an adapted diffusion by Itô's formula. ([10]). The first trivial remark is the following lemma:

Lemma: For each stopping time T , the law of $m_\omega(T)$ starting from $m_0 \in M$ is a Jensen measure for m_0 . (See [6] for a detailed proof.)

As a consequence, if there exists a non trivial adapted diffusion starting from m_0 , then m_0 must have non trivial Jensen measures; let us call the Jensen boundary of A the set of points $m \in M$ such that the only Jensen measure of m is δ_m.

Examples: 1) The algebra of all continuous functions $C(K)$ on a compact space K has no non trivial adapted diffusion.

2) If K is a compact of $\mathbb{C}$, then the Jensen boundary of $R(K)$ (algebra spanned by rational function with no poles on K) is the fine boundary of K , and the only adapted diffusion is the brownian path (or its change of time) stopped at the first exit time of the fine interior of K . (See 3. for this case, and also [5].)

In the same way that analytic structures live on Gleason parts, the adapted diffusions have to live on Jensen parts: let us define the following relation $\sim$ between two points $m,m' \in M$: $m \sim m'$ iff m and m' have two mutually absolutely continuous Jensen measures. This is not an equivalence relation but we call Jensen parts the maximal parts P of M such that the restriction $\sim|_{P\times P}$ is an equivalence relation and that P is one class for $\sim$. With that definition we have the following abstract theorem (which is quite surprising if one thinks that the analoguus result is false with the Gleason part).

Theorem: Let P be a non trivial Jensen part of M for A , $f \in A$ non constant on P . Then for each $m_0 \in P$, $f(m_0)$ is fine interior to $f(M)$ (in the sense of logarithmic potential in $\mathbb{C}$).

In the same way one can prove:

Theorem: Let K be a compact of $\mathbb{C}^n$ which is convex with respect to some function algebra A containing $z_1,\dots,z_n$, and let $p \in K$ be a non Jensen boundary point of K . There exists

$1 \leq i \leq n$ <u>such that</u> $z_i(p)$ <u>is fine interior to</u> $z_i(K)$.

For example, in the case of Stolzenberg example, the projection by the two coordinates functions of $h(K) \setminus K$ has fine interior (but certainly not ordinary interior; in fact, it is this way that Stolzenberg proves that it has no analytic structure).

3. Fine Potential theory

If Δ is some elliptic $2^{\underline{nd}}$ order differential operator on a smooth manifold V , then the fine topology on V associated to Δ is the finest topology such that the subharmonic functions of Δ are continuous with respect to that topology. A nice method to describe this topology is the following one: let $(X_t)_t$ be the diffusion process associated to Δ (i.e. it is a strong Markov process such that the transition functions of this process satisfies the heat equation $\frac{\partial}{\partial t} = \frac{1}{2}\Delta$). Let $A \subset V$ be a Borel set; then $x \in A$ is fine interior to A if the process $(X_t)_t$ stays a little amount of time in A , starting from x before leaving A . By the $0 - 1$ law of Blumenthal Getoor [3] this is the same as

$$P_x(T_A > 0) = 1$$

where T_A is the first exit time of A and P_x is the conditional probability of the process knowing that $X_0 = x$. To see that a Borel function f is finely continuous at x_0 is equivalent to see that the function $t \to f(X_t(\omega))$ is P_{x_0} - a.s. continuous at $t = 0$. (See [3] for the probabilistic interpretation of fine topology.)

<u>Definition</u> (Fuglede [8]): Let U be a fine open set. A function $f: U \to \mathbb{R}$ is finely harmonic, if it is finely continuous and if for every $x_0 \in U$, for every fine neighborhood of x_0 , V such that $\bar{V}^f \subset U$ ($\bar{V}^f$ is the fine closure of V), one has $f(x_0) = \int_{\partial_f V} f(y) d\nu_{x_0}^V(y)$ where $\partial_f V$ is the fine boundary of V and $d\nu_{x_0}^V(y)$ is the swept measure of δ_{x_0} onto $\complement V$.
Here $d\nu_{x_0}^V$ is the the law of the random variable $X_\omega(T_V)$ where T_V is the first exit time of V and $X_\omega(t)$ starts from x_0 at time $t = 0$.

In the same way, one can define finely subharmonic function on U: they are finely u.s.c. functions satisfying the submean property

$$f(x_o) \leq \int_{\partial_f V} f(y)\,dv_{x_o}^V(y) \ .$$

A deep theorem of Fuglede ([8]) asserts that a finely subharmonic function is in fact finely continuous.

In [4] the following results are proved:

(1) Let K be a compact set of V, and f be a continuous function on K which is finely harmonic in the fine interior K' of K. Then f can be uniformly approximated on K by a sequence of ordinary harmonic functions in the neighborhood of K. (See also [9] for a fine local version.)

(2) Let f be a continuous function on K which is finely harmonic in K', fine interior of K, and $(f_n)_n$ a sequence of harmonic functions in the neighborhood of K which converges uniformly on K to f. Then one can write $K' = \bigcup_p V_p$ where V_p are finely open sets in K' such that $(\nabla f_n)_n$ converges in $L^2(V_p, dv)$ towards a vector ∇f. Moreover, for each t one has the following equality

$$f(X_\omega(t \wedge T_K)) = f(x_o) + \int_0^{t \wedge T_K} (\vec{\nabla} f)(X_\omega(s))\,d\vec{X}_\omega(s)$$

in the sense of stochastic integrals and one has also a kind of Sobolev elliptic estimate

$$\|\nabla f\|_{L^2(V_p, dv)} \leq C_p \|f\|_{L^\infty(K, dv)} \ .$$

Remark: In general a finely harmonic function is not continuous in the ordinary topology. But every finely continuous function has the following property: each point x has a fine neighborhood V such that $f|_V$ is continuous for the ordinary topology on V. So the general case for finely harmonic functions is reduced to assertion (1) and (2) finely locally.

4. Algebras R(K) in one complex variable

Let now K be a compact set of $\mathbb{C}$ and let $\mathbb{C}$ be equipped with the ordinary laplacian Δ and the standard brownian motion $b_\omega(t)$. Let $R(K)$ denote the algebra spanned by holomorphic functions in the neighborhood of K and $H(K)$ the vector space spanned by harmonic functions in the neighborhood of K. In [5] the following theorem is proved:

Theorem: (1) The Jensen boundary of $R(K)$ is the fine boundary of K.

(2) At every finely interior point of K, there exists bounded point derivation of every order for $R(K)$.

(3) The Keldych measure (i.e. the swept measure of δ_{x_0} on $\complement K$) is the only Jensen measure for $R(K)$ which is carried by $\partial_f K$.

(4) If $x_0 \in K'$ (fine interior of K as usual) then the Gleason part of x_0 contains at least the fine connected component of K' containing x_0 (and this fine component is finely open because the fine topology is locally connected, see [8]).

In the case of $R(K)$ the structure of Jensen parts is quite simple: the trivial Jensen parts are the points of the fine boundary of K; the other Jensen parts are the fine connected components of the fine interior of K.

Remark 1: In general, $H(K)$ has no bounded point derivation in the fine interior of K. ([4]).

Remark 2: The Harnack principle is true for $R(K)$ in the fine components of K. It is false for $H(K)$. ([4]).

These two remarks (and also some facts about mean approximation) show that the behaviour of finely harmonic function is completely different from that of finely holomorphic function.

In [5], one uses regularity theorems for finely harmonic functions to prove regularity theorems for $R(K)$.

Theorem: (1) There exists a fine open set U, finely dense in K' such that $f^{(p)}|_U$ is finely harmonic in U and satisfies the fine partial differential equation $\frac{\partial f^{(p)}}{\partial \bar{z}}\Big|_U = 0$.

(2) If $(f_n)_n$ is a sequence of holomorphic functions in the neighborhood of K which converges uniformly to $f \in R(K)$, then the sequence $(f_n')_n$ converge finely locally in L^p for $1 \leq p < \infty$ towards f'.

Moreover, one can prove also theorems such like

$$\liminf_{\substack{x \to x_0 \\ \text{finely in } K'}} \left| \frac{f^{(p)}(x) - f^{(p)}(x_0)}{x - x_0} - f^{(p+1)}(x_o) \right| = 0 .$$

One can also obtain results about the structure of extremal Jensen measures of a point x_o : for every finely closed set $\partial_f K \subset A \subset \partial K$, if $\hat{T}_A$ is the first entry time in A starting from x_o , then the law of $b_\omega(\hat{T}_A)$ is an extremal Jensen measure for x_o carried by the boundary ∂K .

5. Construction of holomorphically convex hulls of a class of compact sets in $\mathbb{C}^2$

Let z, w denote the coordinate functions of $\mathbb{C}^2$ and let $D = \{(z,w) \in \mathbb{C}^2 / |z|^2 \leq \phi(|w|^2)\}$ where ϕ is a real analytic function and suppose that D is strictly pseudoconvex. Let S_o denote the hypersurface $|z|^2 = \phi(|w|^2)$.

Let K be a compact set in the w-plane where ϕ does not vanish and suppose K has a regular boundary. Let

$$X_K = \{(z,w) \in \mathbb{C}^2/(z,w) \in S_o, \ w \in K\}$$

and call $u(w)$ the solution of the Dirichlet problem in $\mathring{K}$ with the boundary condition

$$u(w) = -\tfrac{1}{2} \log \phi(|w|^2) \quad \text{on} \quad \partial K .$$

Theorem: The holomorphically convex hull $h(X_K)$ of X_K is:

$$h(X_K) = \{(z,w) \in \mathbb{C}^2/|z|^2 \leq \phi(|w|^2) \text{ and } u(w) + \log|z| \geq 0\}$$

Sketch of proof: Let S_1 be the hypersurface $u(w) + \log|z| = 0$, and S_t , $t \in [0,1]$, the convex combination of S_1 and S_o, so that S_t is strictly pseudoconvex if $0 \leq t < 1$ and S_1 is Levi flat. Taking $t \in [0,1]$ as a local coordinate in $h(X_K)$, one can prove that there exists a 2nd order hyperelliptic operator D_t tangential to S_t such that the holomorphic functions satisfy the heat equation $\frac{\partial f}{\partial t} = D_t f$. D_t is constructed with the Kohn's laplacian ([7]). Then if f is holomorphic near X_K , the analytic continuation of f down to S_1 is given by the heat propagator semi-group $P_t f$ of the preceeding heat equation (see [6] for details).

Remark 1: This provides a new proof of Hartogs' theorem.

Remark 2: This heat equation has been first used by P. Malliavin for the study of boundary value of pluriharmonic functions and the zero sets of Nevanlinna class in strictly pseudoconvex domains, (see [12]).

Remark 3: This method gives also a short proof for Basener's theorem: if K is a compact regular in unit disc Δ such that $\partial\Delta \subset K$ and if

$$X_K = \{(z,w) \in \mathbb{C}^2 / |z| = 1,\ w \in K\} \cup \{(z,w)/z \in K,\ |w| = 1\}$$

then

$$h(X_K) = \{(z,w) \in \Delta^2 / u(z) + u(w) \leq 1\}$$

where u is the solution of the Dirichlet problem in K with given boundary conditions

$$u(z) = \begin{cases} 0 & \text{if } z \in \partial\Delta \\ 1 & \text{if } z \in \partial K \setminus \partial\Delta . \end{cases}$$

6. Holomorphically convex hulls and fine potential

Let us take again the notations of the beginning of 5. but with any compact $K \subset \mathbb{C}$.
Then we have:

Theorem: (1) If $K' = \emptyset$, then $h(X_K) = X_K$.

(2) If $K' \neq \emptyset$, then $h(X_K) \neq X_K$, and $h(X_K) \setminus X_K$ has fine interior for the euclidean potential in $\mathbb{R}^4$.

As corollaries, one obtains information about the 4th dimensional Hausdorff measure of Gleason parts and in the same way as in 3. about the first derivatives $\frac{\partial f}{\partial z}$ and $\frac{\partial f}{\partial w}$ of the function of $H(X_K)$ (i.e. algebra spanned by the functions holomorphic in the neighborhood of X_K).

In fact, one has even stranger results by considering brownian sheets introduced by P. Malliavin for the study of biharmonic functions in the bidisc. ([13]).

Let $b_{\omega_1}(t_1)$ and $b_{\omega_2}(t_2)$ be two independent brownian paths on $\mathbb{C}$; then the brownian sheet in $\mathbb{C}^2$ is defined by the map

$$(t_1, t_2) \to (b_{\omega_1}(t_1),\ b_{\omega_2}(t_2))$$

The set of times $\mathbb{R}^+ \times \mathbb{R}^+$ is ordered by the partial order $(s_1,s_2) \leq (t_1,t_2)$ iff $s_1 \leq t_1$ and $s_2 \leq t_2$. A Borel subset $B \subset \mathbb{C}^2$ is finely open for this biprocess iff for every $(z_0, w_0) \in B$, a.s. in (ω_1, ω_2) there exist $(t_1,t_2) > (0,0)$ such that for every $(s_1,s_2) \leq (t_1,t_2)$, then $(b_{\omega_1}(s_1), b_{\omega_2}(s_2)) \in B$. For example, sets like $U \times V$ (with U and V ordinary fine open sets in $\mathbb{C}$) are finely open for the biprocess. Moreover, if Z is a complex analytic manifold of dimension 1, then $(U \times V) \cap Z$ is a fine open set in Z for the fine topology of Z induced by canonical potential theory on Z.

In the class of examples studied in the preceding theorem, and also in the class of examples studied by Basener, $h(X_K) \setminus X_K$ contains products of fine open sets of $\mathbb{C}$. Using generalized Itô's formulas and area integrals with bitime processes, one can prove the existence of $\frac{\partial^2 f}{\partial z \partial w}$ and Sobolev estimates for these mixed derivatives (see [6] for details of proofs of the assertion in this §).

7. An example of Cole and Wermer

In this conference,[15], J. Wermer introduced the following example (also studied by B. Cole): Let $\Pi = \{(z,\zeta_1 \dots \zeta_n \dots)/|z| \le 1,\ |\zeta_i| \le z\ \forall i\}$, and A the uniform algebra spanned by polynomials in the variables $z,\zeta_1 \dots \zeta_n \dots$ on the compact space Π. Let

$$Y = \{(z,\zeta_1 \dots \zeta_n \dots) \in \Pi / \zeta_i^2 = z - a_i, \forall i\}$$

where $a_1 \dots a_n \dots$ is a dense sequence of points in the unit disc. Then $A|_Y$ is a uniform algebra with maximal ideal space Y and Silov boundary $X = \{(z,\zeta_1 \dots \zeta_n \dots) \in Y/|z| = 1\}$ and J. Wermer proved that there is no analytic structure in $Y \setminus X$.

Nevertheless, it is possible to construct an adapted diffusion in the sense of 2. for this algebra on the spectrum Y. Let us take a point $z_0 \in \Delta$, $z_0 \neq a_i, \forall i$, and then a point y_0 above z_0 in Y:

$$y_0 = (z_0,\zeta_1^0,\dots,\zeta_n^0 \dots) \in Y$$

We want to construct a diffusion process $Y_\omega(t)$ in Y starting from y_0: $Y_\omega(t) = (Z_\omega(t),\zeta_{1,\omega}(t),\dots,\zeta_{n,\omega}(t)\dots)$.

$Z_\omega(t)$ will be the standard brownian path starting from z_0 at time $t = 0$ stopped at the first exit time of the unit disc. $\zeta_{n,\omega}(t)$ will be the lifted path of $Z_\omega(t)$ starting from $\zeta_{n,0}$ at time $t = 0$ in the covering space defined by the Riemann surface of $\zeta_n^2 = z - a_n$; let us just remark that there is no difficulty of construction of all these lifted paths for all n, because the brownian path $Z_\omega(t)$ downstairs a.s. never hits the ramification set $(a_1 \dots a_n \dots)$ of Y because this set is countable, hence of capacity 0 (and in fact that construction would work with any ramification set of 0 capacity instead of a countable set).

Now it is clear that the polynomials computed along these paths give conformal martingales, so that we have constructed an adapted diffusion.

REFERENCES.

[1] R. Basener: "Generalized Shilov Boundary",(preprint, Yale University).

[2] R. Basener: "On Some Rationally Convex Hulls", (Trans. Am. Math. Soc., August 1973).
"Rationally Convex Hulls and Potential Theory", (Preprint, Yale University, (1974).

[3] R. Blumenthal, R. Getoor: "Markov Processes and Potential Theory", Academic Press, 1968.

[4] A. Debiard, B. Gaveau: "Différentiabilité des fonctions finement harmoniques". A paraître aux Inventiones Mathematicae, 1975.

[5] A. Debiard, B. Gaveau: "Potentiel fin et algèbres de fonctions analytiques" I Journal of Functional Analysis, July 1974, II ibidem November 1974.

[6] A. Debiard, B. Gaveau: "Potentiel fin et enveloppes d'holomorphie" III to appear in Journal of Functional Analysis, January 1976.

[7] G. Folland, J.J. Kohn: "The $\bar{\partial}$ - Neumann Problem", Annals of Math. Studies, Princeton 1972.

[8] B. Fuglede: "Fine Connectivity and Finely Harmonic Functions", Actes Congrès International, Nice 1970. (Gauthier Villars editeur)

B. Fuglede: "Finely Harmonic Functions", Springer Lecture Notes in Maths.

[9] B. Fuglede: "Fonctions harmoniques et fonctions finement harmoniques", Annales de l'Institut Fourier, T 24, 1974, p. 77.

[10] Mac Kean: "Stochastic Integrals", Academic Press 1969.

[11] P. Malliavin: "Comportement d'une fonction analytique et plusieurs variables á la frontière distinguée", Comptes Rendus Acad. Sc. Paris, fér 1969.

[12] P. Malliavin: "Equation de la chaleur associeé à un fonction plurisousharmonique ", (preprintá paraître).

[13] P. Malliavin: "Processus á temps bidimensionnal dans le bidisque".

[14] G. Stolzenberg: "A Hull with no Analytic Structure", Journal of Maths. and Mechanics 1964.

[15] J. Wermer, this conference.

BOUNDED POINT EVALUATIONS AND APPROXIMATION IN L^p BY ANALYTIC FUNCTIONS.

Claes Fernstrøm

Let us consider the complex plane $\mathbb{C}$. We shall assume all the time that E is a compact set without interior. Let $\eta(E) = \{f\ ;\ f$ is analytic in a neighbourhood of $E\}$ and let $\|f\|_p^p = \int_E |f(z)|^p dm(z)$, where $1 \leq p < \infty$ and where m denotes the Lebesgue measure.

Definition 1. z_0, $z_0 \in E$, is a bounded point evaluation for $\eta(E) \subset L^p(E)$ if there is a constant C such that $|f(z_0)| \leq C\|f\|_p$ for all f, $f \in \eta(E)$.

The following theorem is due to Brennan [1].

Theorem 1. Assume that $p \neq 2$. Then $\eta(E)$ is dense in $L^p(E)$ if and only if almost every point of E fails to be a bounded point evaluation for $\eta(E) \subset L^p(E)$.

The purpose of this paper is to prove the following theorem.

Theorem 2. There is E such that

(i) there is no bounded point evaluations for $\eta(E) \subset L^2(E)$

(ii) $\eta(E)$ is not dense in $L^2(E)$.

In order to prove theorem 2 we shall rewrite theorem 2 in terms of certain capacities. Let $G(x)$, $x \in R^2$, be the Bessel kernel, defined as the Fourier transform of $(1+|x|^2)^{-\frac{1}{2}}$. Assume that q is a number such that $1 < q < \infty$ and $\frac{1}{p} + \frac{1}{q} = 1$.

Definition 2. Let A be a set in $\mathbb{R}^2$. Then $C_q(A) = \inf \int_{\mathbb{R}^2} |f(x)|^q dm(x)$, where the infimum should be taken over all f, $f \in L^q(R^2)$, such that $f \geq 0$ and $\int_{\mathbb{R}^2} G(x-y)f(y)dm(y) \geq 1$ for all x, $x \in A$.

This set function has been studied e.g. by Meyers [2]. The following theorem is due to Hedberg [3].

Theorem 3. Let $2 \leq p < \infty$. Then the following are equivalent.

(a) $\eta(E)$ is dense in $L^p(E)$.

(b) $C_q(\Omega \setminus E) = C_q(\Omega)$ for all open Ω.

(c) $\overline{\lim_{\delta\to 0}} \dfrac{C_q(B_z(\delta)\setminus E)}{\delta^2} > 0$ for almost all z, where $B_z(\delta) = \{\zeta \in \mathbb{C}\ ;\ |\zeta - z| < \delta\}$.

Theorem 4. Let $2 \leq p < \infty$. Then z_0 is a bounded point evaluation for $\eta(E) \subset L^p(E)$ if and only if $\sum_{k=1}^{\infty} 2^{kq} C_q(E \setminus A_k(z_0)) < \infty$, where $A_k(z_0) = \{z \in \mathbb{C}\ ;\ 2^{-k-1} < |z-z_0| \leq 2^{-k}\}$.

Theorem 4 has been proved by Hedberg for $p > 2$ [4]. He has also proved in [4] that the condition in Theorem 4 is necessary for $p = 2$. We shall not need that the condition is sufficient for $p = 2$, therefore we are not going to prove it here.
Now one sees immediately that Theorem 2 can be formulated in terms of capacities.

Theorem 2'. There is E such that

(i) $\sum_{k=1}^{\infty} 2^{2k} C_2(A_k(z) \setminus E) = \infty$ for all z, $z \in \mathbb{C}$

(ii) $C_2(B_0(\tfrac{1}{2}) \setminus E) < C_2(B_0(\tfrac{1}{2}))$.

Proof. (We do not intend to give all the details).
There are constants F_1 and F_2 such that

$$\frac{F_1}{\log \frac{1}{\delta}} \leq C_2(B_z(\delta)) \leq \frac{F_2}{\log \frac{1}{\delta}} \quad \text{for all } \delta, \quad \delta \leq \delta_0 < 1.$$

Choose α, $\alpha \geq 1$ such that

$$\sum_{n=1}^{\infty} \frac{F_2}{\alpha n^2} < C_2(B_0(\tfrac{1}{2})) .$$

Let A_0 be the closed unit square with centre in origin. Cover A_0 with 4^n squares with side 2^{-n}. Call the squares $A_n^{(i)}$, $i = 1, 2, \ldots, 4^n$. In every $A_n^{(i)}$ put an open disc $B_n^{(i)}$, $i = 1, 2, \ldots, 4^n$, such that $B_n^{(i)}$ and $A_n^{(i)}$ have the same centre and such that the radius of $B_n^{(i)}$ is $e^{-\alpha 4^n n^2}$.
Repeat the construction for all n, $n \geq 1$.

Put $E = A_0 \setminus \left(\bigcup_{n=1}^{\infty} \bigcup_{i=1}^{4^n} B_n^{(i)} \right)$

If one uses the subadditivity of C_2, one can easily prove (ii).
In order to prove (i) it is enough to prove that

$$C_2(A_n^{(i)} \setminus E) \geq \frac{F_1}{16\alpha n 4^n} \quad \text{for all } n, \; n > n_0 \quad \ldots\ldots\ldots\ldots (1)$$

Let us consider all $B_k^{(j)}$, $n \leq k \leq 2n$, such that $B_k^{(j)} \subset A_n^{(i)}$. We get 4^{ℓ} discs with radius $e^{-\alpha 4^{n+\ell}(n+\ell)^2}$, $0 \leq \ell \leq n$. Call the discs $D_n^{(\nu)}$, $\nu = 1, 2, \ldots, \frac{4^{n+1}-1}{3}$.

$$\therefore \frac{1}{4^n} \sum_{j=n}^{2n} \frac{F_1}{\alpha_j^2} \leq \sum_{\nu} C_2(D_n^{(\nu)}) \leq \frac{1}{4^n} \sum_{j=n}^{2n} \frac{F_2}{\alpha_j^2} \quad \ldots\ldots\ldots\ldots (2)$$

Put $D_n = \bigcup_{\nu} D_n^{(\nu)}$.

Choose n_1 such that

$$\text{Dist}(D_n^{(\nu)}, D_n^{(\mu)}) \geq 4e^{-4^n} \quad \text{for all } n, \; n \geq n_1 \text{ and } \nu \neq \mu \quad \ldots\ldots (3)$$

Choose $f_n \in L^2(R^2)$ such that $f_n \geq 0$,

$$\int G(x-y) f_n(y)\, dm(y) \geq 1 \quad \text{for all } x, \; x \in D_n \quad \ldots\ldots\ldots\ldots (4)$$

$$\text{and} \int |f_n(x)|^2 dm(x) \leq 2C_2(D_n) \leq 2 \sum_{\nu} C_2(D_n^{(\nu)}) \quad \ldots\ldots\ldots\ldots (5)$$

(2) and (5) now give

$$x \in D_n^{(\nu)} \Rightarrow \int_{\text{Dist}(y, D_n^{(\nu)}) \geq e^{-4^n}} G(x-y) f_n(y)\, dm(y)$$

$$\leq \left[\int_{|y| \geq e^{-4^n}} G(y)^2 dm(y) \int |f_n(y)|^2 dm(y) \right]^{\frac{1}{2}}$$

$$\leq \text{const.} \left[4^n \cdot \frac{1}{4^n} \sum_{j=n}^{2n} \frac{F_2}{j^2} \right]^{\frac{1}{2}} < \frac{1}{2} \quad \text{if } n \text{ is big enough.}$$

$\therefore$ We can choose $n_0 (n_0 \geq n_1)$ such that

$$\int_{\text{Dist}(y,D_n^{(\nu)}) \leq e^{-4^n}} G(x-y)2 f_n(y)dm(y) \geq 1 \quad \text{for all} \quad x, x \in D_n^{(\nu)} \quad \text{and}$$

for all $n, n \geq n_0$ .... (6)

Let in the following all n be bigger than n_0. Choose $g_n^{(\nu)}$ such that

$$g_n^{(\nu)}(x) = \begin{cases} 2 f_n(x) & \text{if } \text{Dist}(D_n^{(\nu)},x) \leq e^{-4^n} \\ 0 & \text{if } \text{Dist}(D_n^{(\nu)},x) > e^{-4^n} \end{cases}$$

If we use (6), we get that

$$x \in D_n^{(\nu)} \Rightarrow \int G(x-y)g_n^{(\nu)}(y)dm(y) \geq 1$$

$$\therefore \int |g_n^{(\nu)}(x)|^2 dm(x) \geq C_2(D_n^{(\nu)})$$

(2), (3) and (5) now give

$$C_2(D_n) \geq \frac{1}{2} \int |f_n(x)|^2 dm(x) \geq \frac{1}{8} \sum_\nu \int |g_n^{(\nu)}(x)|^2 dm(x) \geq \frac{1}{8} \sum_\nu C_2(D_n^{(\nu)})$$

$$\geq \frac{1}{8} \cdot \frac{1}{4^n} \sum_{j=n}^{2n} \frac{F_1}{\alpha j^2} \geq \frac{F_1}{16\alpha 4^n n}$$

$$\therefore \quad C_2(A_n^{(i)} \setminus E) \geq \frac{F_1}{16\alpha n 4^n} \quad \text{for all } n, n > n_0, \text{ which is (1).}$$

The results above can be generalized to solutions of elliptic partial differential equations. These results will appear.

REFERENCES

[1] J. Brennan: Invariant subspaces and rational approximation. J. Functional Analysis 7, 285 - 310 (1971).

[2] N.G. Meyers: A theory of capacities for potentials of functions in Lebesgue classes. Math. Scand. 26, 255 - 292 (1970).

[3] L.I. Hedberg: Non-linear potentials and approximation in the mean by analytic functions. Math. Z. 129, 299 - 319 (1972).

[4] L.I. Hedberg: Bounded point evaluations and capacity. J. Functional Analysis 10, 269 - 280 (1972).

HARTOGS SERIES, HARTOGS FUNCTIONS AND JENSEN MEASURES

T.W. Gamelin

INTRODUCTION

A Hartogs series is a series of the form

(0.1) $$\sum f_j \zeta^j,$$

where ζ is a complex variable, and the f_j depend on other parameters. These series arise naturally in connection with analytic functions on subsets of $\mathbb{C}^{n+1}$ that are "circled" in one variable, that is, that are invariant under the transformations

(0.2) $$(z,\zeta) \to (z, e^{i\theta}\zeta), \qquad z \in \mathbb{C}^n,\ \zeta \in \mathbb{C},\ 0 \le \theta \le 2\pi.$$

An analytic function on such a set can be expanded in a Hartogs series (0.1), where the coefficients f_j depends analytically on the parameter z. When one studies the dependence of the annulus of convergence of the series on the parameter, one is led to Hartogs functions, and through duality one arrives at Jensen measures. We begin by discussing Jensen measures and Hartogs functions in the context of a uniform algebra, and we specialize later to invariant subsets of $\mathbb{C}^{n+1}$.

§1. Jensen Measures and Hartogs Functions

Let X be a compact space, let A be a uniform algebra on X, and let M_A be the maximal ideal space of A. Recall that a Jensen measure on X for $\varphi \in M_A$ is a probability measure σ on X that satisfies

$$\log|f(\varphi)| \le \int \log|f|\,d\sigma, \qquad f \in A.$$

The Jensen measures for φ form a weak-star compact, convex set of measures on X.

The family of Hartogs functions on M_A is the smallest family $\mathcal{H}$ of functions from M_A to $[-\infty,+\infty)$ such that

(1.1) $c\log|f| \in \mathcal{H}$ whenever $f \in A$ and $c > 0$, and

(1.2) $\limsup\limits_{k\to\infty} v_k$ belongs to $\mathcal{H}$ whenever $\{v_k\}_{k=1}^{\infty}$ is a sequence in $\mathcal{H}$ that is bounded above.

Any Hartogs function is a Baire function that is bounded above. The family of Hartogs functions is closed under addition, and under multiplication by positive scalars, so that $\mathcal{H}$ forms a convex cone. The maximum of any two Hartogs functions is again a Hartogs function.

The following theorem exhibits the duality between Hartogs functions and Jensen measures.

<u>Theorem</u> 1.(<u>D. A. Edwards</u>): Let $\varphi \in M_A$, and let $u : X \to (-\infty, +\infty]$ be lower semi-continuous. Then the following are equal:

$$(1.3) \qquad \sup\left\{\frac{\log|g(\varphi)|}{m} : g \in A, m \in Z_+, \frac{\log|g|}{m} < u \text{ on } X\right\}$$

$$(1.4) \qquad \sup\{v(\varphi) : v \in \mathcal{H}, v < u \text{ on } X\}$$

$$(1.5) \qquad \inf\{\textstyle\int u\, d\sigma : \sigma \text{ is a Jensen measure on } X \text{ for } \varphi\}.$$

<u>Proof</u>. Here Z_+ is the set of positive integers. The inequality (1.3) $\le$ (1.4) follows from the definition of $\mathcal{H}$, while the inequality (1.4) $\le$ (1.5) follows from Fatou's Lemma. The proof that (1.5) $\le$ (1.3), in the case that u is continuous, runs in outline as follows.

Let C be the set of $w \in C_R(X)$ for which there are $f \in A$ and $m \in Z_+$ satisfying $f(\varphi) = 1$ and $(\log|f|)/m < w$ on X. Then C is a convex cone in $C_R(X)$ which contains the positive functions. It is easy to check that the Jensen measures for φ consist of precisely those measures of unit norm which are nonnegative on C.

Let b be the infimum appearing in (1.5), and let $\varepsilon > 0$. Then $\int (u - b + \varepsilon) d\sigma \ge \varepsilon$ for all Jensen measures σ for φ. In view of the Separation Theorem for Convex Sets, we see that $u - b + \varepsilon$ belongs to the cone C. Choose $f \in A$ and $m \in Z_+$ such that $f(\varphi) = 1$ and $(\log|f|)/m < u - b + \varepsilon$ on X, and set $g = e^{mb-m\varepsilon} f$. Then $(\log|g|)/m < u$ on X, while $(\log|g(\varphi)|)/m = b - \varepsilon$. It follows that (1.3) $\ge$ (1.5).

The theorem is now proved for continuous u. To pass to the semi-continuous case, one simply invokes the Minimax Theorem. Indeed, (1.5) is equal to

$$\inf_{\sigma} \sup_{v \in C_R, v<u} \int v\, d\sigma = \sup_{v \in C_R, v<u} \inf_{\sigma} \int v\, d\sigma$$

$$= \sup_{v \in C_R, v<u} \sup\left\{\frac{\log|g(\varphi)|}{m} : g \in A, m \in Z_+, \frac{\log|g|}{m} < u \text{ on } X\right\},$$

and this latter quantity coincides with (1.3).

Q.E.D.

§2. <u>Uniform Algebras Generated by Hartogs Series</u>

Let $R : X \to [0,\infty)$ be upper semi-continuous, and define a compact subset Y of $X \times \mathbb{C}$ by

$$(2.1) \qquad Y = \{(x,\zeta) : |\zeta| \le R(x)\}.$$

Let B be the uniform closure in $C(Y)$ of Hartogs polynomials of the form

$$F(x,\zeta) = \sum_{j=0}^{N} f_j(x)\zeta^j, \qquad N \geq 0,\ f_0,\ldots,f_N \in A.$$

Then B is a uniform algebra on Y, and it is clear that the maximal ideal space of B has the form

(2.2) $$M_B = \{(\varphi,\zeta) \in M_A \times \mathbf{C} : |\zeta| \leq S(\varphi)\},$$

for some function $S : M_A \to [0,\infty)$. Since M_B is compact, S is upper semicontinuous.

Each $F \in B$ depends analytically on ζ in the disc $\{(\varphi,\zeta) : |\zeta| < S(\varphi)\}$, so that F has a Hartogs series expansion

(2.3) $$F(\varphi,\zeta) = \sum_{j=0}^{\infty} f_j(\varphi)\zeta^j, \qquad |\zeta| < S(\varphi).$$

Note that the functions f_j need not be bounded on the set $\{S > 0\}$.

Let $R_F(\varphi)$ denote the radius of convergence of the series (2.3). Then $R_F \geq R$ on X, while $R_F \geq S$ on M_A. In fact,

(2.4) $$S(\varphi) = \inf\{R_F(\varphi) : F \in B\},$$

since the quantity on the right is the radius of the smallest disc on which all functions in B can be defined in a natural manner to be analytic. Hartogs functions enter the picture through the expression for the radius of convergence of a power series in terms of the coefficients:

(2.5) $$-\log R_F(\varphi) = \limsup_{j\to\infty} \frac{\log|f_j(\varphi)|}{j}.$$

This shows that $-\log R_F$ is a Hartogs function on M_A just as soon as the f_j's do not grow in norm too rapidly. From (2.4), we expect $-\log S$ to be an upper envelope of Hartogs functions. Indeed, we have the following description of S.

<u>Theorem</u> 2. The function S defined by (2.2) is given by

(2.6) $$S(\varphi) = \inf\{|f(\varphi)|^{-1/m} : f \in A,\ m \in Z_+,\ |f|^{1/m} R < 1 \text{ on } X\}.$$

In particular, the following are each equal to $-\log S(\varphi)$:

(2.7) $$\sup\left\{\frac{\log|f(\varphi)|}{m} : f \in A,\ m \in Z_+,\ \frac{\log|f|}{m} < -\log R \text{ on } X\right\},$$

(2.8) $$\sup\{v(\varphi) : v \in \mathcal{H},\ v < -\log R \text{ on } X\},$$

(2.9) $$\inf\{-\int \log R\, d\sigma : \sigma \text{ is a Jensen measure on } X \text{ for } \varphi\}.$$

Proof. We follow a line of proof suggested by N. Sibony.

In view of Theorem 1, it suffices to establish (2.6). Let β denote the infimum that appears in (2.6).

Suppose $f \in A$ and $m \in Z_+$ satisfy $|f|^{1/m} R < 1$ on X. Then $|f|^{1/m} |\zeta| < 1$ on Y, so that the series $\sum_{j=0}^{\infty} f^j \zeta^{mj}$ converges uniformly on Y. However, the series diverges for $\zeta = |f(\varphi)|^{-1/m}$. It follows that $S(\varphi) \leq |f(\varphi)|^{-1/m}$, and hence $S(\varphi) \leq \beta$.

For the reverse inequality, we assume that $R \geq \varepsilon > 0$; the general case is handled by replacing R by $R+\varepsilon$ and letting ε decrease to 0.

Let $F \in B$. In the case at hand, the coefficients of (2.3) are estimated by

$$|f_j(x)| \leq R(x)^{-j} \|F\|_Y \leq \varepsilon^{-j} \|F\|_Y. \tag{2.10}$$

In particular, the f_j depend continuously on F, so that the f_j belong to A. The first inequality in (2.10) shows that $|f_j|^{1/j} \|F\|^{-1/j} R \leq 1$ on X. From the definition of β, we obtain $|f_j(\varphi)|^{-1/j} \|F\|^{1/j} \geq \beta$, or

$$|f_j(\varphi)| \beta^j \leq \|F\|_Y, \qquad F \in B.$$

It follows that the series $\sum f_j \zeta^j$ converges whenever $|\zeta| < \beta$, so that $R_F(\varphi) \geq \beta$ for all $F \in B$. By (2.4), $S(\varphi) \geq \beta$.

Q.E.D.

The question arises as to whether the function $-\log S$ appearing in Theorem 2 is itself a Hartogs function. The following theorem shows that the answer is affirmative providing X is metrizable and $\log S$ is bounded.

Theorem 3. Let R and S be as in Theorem 2. Suppose that S is a pointwise limit of a decreasing sequence of continuous functions. Then there is a sequence $\{f_j\}_{j=0}^{\infty}$ in A such that the Hartogs series $F = \sum f_j \zeta^j$ converges uniformly and absolutely on M_B, and such that $R_F = S$. In particular, if also S is bounded away from zero, then $-\log S$ is a Hartogs function.

Proof. Let u be a continuous function on M_A such that $u > S$. The identity (2.6) shows that for each $\varphi_0 \in M_A$, there exist $f \in A$ and $m \in Z_+$ such that $|f|^{1/m} R < 1$ on X, while $|f(\varphi_0)|^{-1/m} < u(\varphi_0)$. Then $|f\zeta^m| < 1$ on M_B, while $[\log|f(\varphi_0)|]/m > -\log u(\varphi_0)$. Since u is continuous, this latter identity persists in a neighborhood of φ_0. Using the compactness of M_A and the continuity of u, we can therefore find a finite family of $f_j \in A$ and $m_j \in Z_+$ such that

$$(2.11)\qquad |f_j \zeta^{m_j}| < 1 \quad \text{on } M_B,$$

$$(2.12)\qquad \max_j \frac{\log|f_j(\varphi)|}{m_j} > -\log u(\varphi), \qquad \varphi \in M_A.$$

By approximating S from above by a sequence of such functions u, and by taking the aggregate of the corresponding f_j's, we obtain sequences $\{f_j\}$ in A and $\{m_j\}$ in Z_+ such that (2.11) is valid, and also

$$(2.13)\qquad \limsup_{j\to\infty} \frac{\log|f_j(\varphi)|}{m_j} \geq -\log S(\varphi), \qquad \varphi \in M_A.$$

Replacing f_j by $f_j^{k_j}$ and m_j by $m_j k_j$ for $k_j \in Z_+$ large, we can assume that the integers m_j are distinct, that $|f_j \zeta^{m_j}| < 1/j^2$ on M_B, and that (2.13) remains valid. Then $\sum f_j \zeta^{m_j}$ is the Hartogs series with the desired properties.

Q.E.D.

§3. Uniform Algebras Generated by Hartogs-Laurent Series

The maximal ideal space of an algebra generated by Hartogs series of the form

$$(3.1)\qquad F(x,\zeta) = \sum_{j=-N}^{N} f_j(x)\zeta^j, \qquad f_j \in A,\ -N \leq j \leq N,$$

can be treated in the same manner. In this case, define

$$Y = \{(x,\zeta) \in X \times \mathbf{C} : Q(x) \leq |\zeta| \leq R(x)\},$$

where Q and R are real-valued functions on X, $0 < Q \leq R$, Q is lower semi-continuous, and R is upper semi-continuous. Then R is bounded, Q is bounded away from zero, and Y is compact.

Theorem 4. Let Y be as above, and let B be the uniform closure of the functions (3.1) in $C(Y)$. Then the maximal ideal space of B has the form

$$(3.2)\qquad M_B = \{(\varphi,\zeta) \in M_A \times \mathbf{C} : P(\varphi) \leq |\zeta| \leq S(\varphi)\},$$

where P and S are the semi-continuous functions on M_A given by

$$(3.3)\qquad \log P(\varphi) = \inf\{\textstyle\int \log Q\, d\sigma : \sigma \text{ a Jensen measure on } X \text{ for } \varphi\},$$

$$(3.4)\qquad \log S(\varphi) = \sup\{\textstyle\int \log R\, d\sigma : \sigma \text{ a Jensen measure on } X \text{ for } \varphi\}.$$

Proof. Suppose first that $R > Q + \varepsilon$ for some $\varepsilon > 0$. Each $F \in B$ can then be expanded in a Hartogs-Laurent series

$$F(x,\zeta) = \sum_{j=-\infty}^{\infty} f_j(x)\zeta^j, \qquad x \in X,\ Q(x) < |\zeta| < R(x).$$

Again the f_j depend continuously on F, so that the f_j belong to A. Now the proof of Theorem 2 shows that the series

$$\sum_{j=0}^{\infty} f_j(\varphi)\zeta^j, \qquad \varphi \in M_A,$$

converges providing $|\zeta| < S(\varphi)$, where S is given by the formulae (2.6)-(2.9), that is, by (3.4). Similarly, the series $\sum_{j=-\infty}^{-1} f_j(\varphi)\zeta^j$ converges providing $|\zeta| > P(\varphi)$, where P is given by (3.3). We conclude that M_B includes the set described in (3.2), and the proof of Theorem 2 shows that in fact M_B is given precisely by (3.2).

The general case is handled by replacing R by $R + \varepsilon$, and letting ε decrease to zero.

Q.E.D.

We mention in passing that the algebra B of Theorem 4 is a proper subalgebra of $C(Y)$ just as soon as A is a proper subalgebra of $C(X)$. Indeed, for $F \in B$ and $\varphi \in M_A$, define $F_0(\varphi)$ to be the average of F over the circle $\{(\varphi,\zeta) : |\zeta| = S(\varphi)\}$. Then F_0 depends continuously on F, and F_0 coincides with the coefficient f_0 whenever F has the form (3.1). It follows that F_0 belongs to A for all $F \in B$. In particular, the assertion above is established.

One interesting special case occurs when $Q = R$ is continuous, while every $\varphi \in M_A$ has a unique Jensen measure σ_φ on X. In this case, the expressions (3.3) and (3.4) coincide, so that $P = S$, and

$$M_B = \{(\varphi,\zeta) : \log|\zeta| = \int_X \log R(x)d\sigma_\varphi(x),\ \varphi \in M_A\}.$$

Topologically M_B is the product of M_A and a circle.

A striking example, obtained in this fashion, is R. Basener's example of a uniform algebra B such that $B \neq C(M_B)$, while every point of M_B is a peak point for B. (A more complicated example with this property had been obtained earlier by B. Cole.) Indeed, let K be a compact subset of the closed unit disc in $\mathbf{C}$ such that $R(K) \neq C(K)$, while every point of K has a unique Jensen measure on K - McKissick's Swiss cheese is such a set. Define $Q(z) = R(z) = (1 - |z|^2)^{1/2}$. Then the construction above, with $A = R(K)$, produces an algebra B on a compact subset Y of $\mathbf{C}^2$ such that $M_B = Y$, while $B \neq C(Y)$. Since Y lies on the unit sphere of $\mathbf{C}^2$, each $(z,\zeta) \in Y$ is a peak point for a linear function of z and ζ, and B is the required example.

The example mentioned in his talk by B. Gaveau also falls in this class of algebras.

§4. Plurisubharmonic Functions and Bremermann's Theorem

As mentioned in the introduction, we can obtain from this framework a description of the polynomial hulls of compact subsets of C^{n+1} that are invariant under the transformations (0.2), that is, that are circled in the last variable. Indeed, suppose K is such a compact set. Let X be the projection of K onto the plane of the first n coordinates, and define

$$R(z) = \sup\{|\zeta| : (z,\zeta) \in K\}, \qquad z \in X. \tag{4.1}$$

Define Y as before, by (2.1). Then $Y \subseteq K$, and the polynomial hulls of Y and K evidently coincide. They have the form

$$\hat{Y} = \hat{K} = \{(z,\zeta) \in C^{n+1} : z \in \hat{X}, |\zeta| \leq S(z)\}, \tag{4.2}$$

for some appropriate function S on $\hat{X}$. The function S is determined from R by applying Theorem 2 to the algebra $P(X)$. In this case, the following version of a theorem of H. Bremermann allows us to replace the family of Hartogs functions by the family of plurisubharmonic functions.

Theorem 5 (Bremermann). Let X be a compact subset of C^n. If u is defined and plurisubharmonic in a neighborhood of the polynomial hull $\hat{X}$ of X, then u is a Hartogs function on $\hat{X}$ with respect to the algebra $P(X)$.

The proof is based on the following lemma.

Lemma. Let D be a domain of holomorphy in C^n, and let v be a (upper semi-continuous) plurisubharmonic function on D. Then the domain

$$D^* = \{(z,\zeta) \in C^{n+1} : z \in D, |\zeta| < e^{-v(z)}\}$$

is a domain of holomorphy in C^{n+1}. Furthermore, if D is a Runge domain, then D^* is a Runge domain.

Proof. We assume that v is smooth. The general case is obtained from the smooth case by using standard techniques.

On account of the hypothesis on D, every point of ∂D^* lying over a point of ∂D is a singular boundary point for some holomorphic function on D^* that is independent of the variable ζ. On the other hand, by checking the Hessian of a suitable defining function, we deduce from the plurisubharmonicity of u that the portion of ∂D^* lying above D is pseudoconvex. Hence D^* is a domain of holomorphy.

Every holomorphic function F on D^* is approximable uniformly on compacta by its Hartogs expansions $F(z,\zeta) = \sum f_j(z)\zeta^j$, and the f_j depend holomorphically on z. If D is a Runge domain, then the f_j can be approximated uniformly on compacta by polynomials, so that F is approximable uniformly on compacta by polynomials, and D^* is also a Runge domain.

Q.E.D.

It is not difficult to produce an analytic function F on D^* whose Hartogs series has radius of convergence $R_F = v$. We proceed though to the proof of Theorem 5, which boils down to a related question.

<u>Proof of Theorem</u> 5. We can assume that u is smooth. Let D be a Runge domain containing $\hat{X}$ such that u is smooth and plurisubharmonic on D. Let $\varepsilon > 0$, and define

$$D^* = \{(z,\zeta) \in \mathbb{C}^{n+1} : z \in D,\ |\zeta| < e^{-u(z)+\varepsilon}\},$$

$$X^* = \{(z,\zeta) \in \mathbb{C}^{n+1} : z \in \hat{X},\ |\zeta| \leq e^{-u(z)}\}.$$

By the Lemma, D^* is a domain of holomorphy and a Runge domain. Since X^* is a compact subset of D^*, so is the polynomial hull $\widehat{X^*}$ of X^*.

Let $z_0 \in \hat{X}$. Since $\widehat{X^*}$ is a compact subset of D^*, there is a polynomial $F(z,\zeta)$ such that $|F| < 1/2$ on X^*, while $F(z_0,\zeta_0) = 1$ for some point $(z_0,\zeta_0) \in D^*$. We claim that for each z_1 near z_0, there exists ζ_1 such that $(z_1,\zeta_1) \in D^*$, while $F(z_1,\zeta_1) = 1$. Indeed, $F(z_0,\zeta)$ is nonconstant as a function of ζ, so that the winding number of $F(z_0,\zeta) - 1$ around a small circle $\{|\zeta-\zeta_0| = \delta\}$ is defined and positive for all z_1 near z_0, and $F(z_1,\zeta)$ assumes the value 1 inside the circle.

Since $\hat{X}$ is compact, there exist a finite number of polynomials $F_1,\dots,F_k$ such that $|F_j| < 1/2$ on X^*; while for each $z \in \hat{X}$, there exists ζ satisfying $|\zeta| < e^{-u(z)+\varepsilon}$, and $F_j(z,\zeta) = 1$ for some index j. The function $G = 1/[(1-F_1)\dots(1-F_k)]$ is analytic in a neighborhood of X^*, so its Hartogs series has radius of convergence

$$R_G(z) \geq e^{-u(z)}, \qquad z \in \hat{X}. \tag{4.3}$$

On the other hand, for each fixed $z \in \hat{X}$, $G(z,\zeta)$ has a pole at some point ζ satisfying $|\zeta| < e^{-u(z)+\varepsilon}$. Hence

$$R_G(z) \leq e^{-u(z)+\varepsilon}, \qquad z \in \hat{X}. \tag{4.4}$$

From (4.3) and (4.4), we have the uniform estimate $|u(z) + \log R_G(z)| < \varepsilon$, $z \in \hat{X}$. Since $-\log R_G$ is a Hartogs function with respect to $P(\hat{X})$, u is also a Hartogs function.

Q.E D.

Now the family of plurisubharmonic functions is sandwiched between the family of Hartogs functions and the family of functions of the form $c \log|f|$, where $c > 0$ and f is a polynomial. Consequently Theorems 1 and 2 can be rephrased as follows.

Theorem 6. Let K be a compact subset of C^{n+1} that is invariant under the transformations (0.2). Let X be the projection of K into C^n, and define R by (4.1). Then the polynomial hull $\hat{K}$ of K is given by (4.2), where $-\log S$ is the upper envelope of the family of functions u which are plurisubharmonic on a neighborhood of $\hat{X}$ and which satisfy $u < -\log R$ on X.

Theorem 6 can be considered classical. It is a simple consequence of an old result describing the envelopes of holomorphy of certain domains in C^{n+1} of the form

$$D_R = \{(z,\zeta) \in C^{n+1} : z \in D, |\zeta| < e^{-R(z)}\}.$$

The theorem asserts that if D is a domain of holomorphy in C^n, and if $R: D \to [-\infty, +\infty)$ is lower semi-continuous, then the envelope of holomorphy of D_R has the same form, say D_S, where $-\log S$ is the upper envelope of the family of plurisubharmonic functions u on D that satisfy $u < -\log R$. This latter theorem can also be deduced easily from the Lemma above. For details, see the book of V. S. Vladimirov listed in the bibliography, and Bremermann [1].

More recently, N. Sibony has obtained a related result, determining the $H^\infty(D)$-hulls of certain domains in C^{n+1} that are circled in the last variable. The description runs along the same lines as the results we have been describing, except that the family of plurisubharmonic functions must be replaced by an appropriate family of Hartogs functions, namely, the Hartogs family generated by $\log|f|$, $f \in H^\infty(D)$. In connection with this problem, Sibony has noted the following remarkable example.

Example (Sibony). There exists a proper subdomain D of the unit bidisc $\Delta\times\Delta$ in C^2, such that D is the interior of its closure, D is domain of holomorphy, and every bounded analytic function on D extends to be a bounded analytic function on $\Delta\times\Delta$.

Since the construction of the example is related to the circle of ideas under discussion, we give the details.

Let $\{a_j\}_{j=1}^\infty$ be a sequence of points in the open unit disc Δ that clusters on $\partial\Delta$, and that satisfies

$$\sup_{1\leq j<\infty} |f(a_j)| = \sup\{|f(z)| : z \in \Delta\}, \qquad f \in H^\infty(\Delta). \tag{4.5}$$

Define

$$u(z) = \prod_{j=1}^{\infty} \left|\frac{z-a_j}{2}\right|^{b_j}, \qquad z \in \Delta,$$

where the sequence $\{b_j\}$ decreases so rapidly to zero that the product converges uniformly on compact subsets of Δ. Then u is subharmonic on Δ, $0 \leq u < 1$, and $u(a_j) = 0$, $j \geq 1$. Define

$$D = \{(z,\zeta) : z \in \Delta,\ |\zeta| < e^{-u(z)}\}.$$

Then D is a proper subdomain of $\Delta \times \Delta$ which coincides with the interior of its closure. By the Lemma, D is a domain of holomorphy.

Let $F \in H^{\infty}(D)$, and let $\sum f_j(z)\zeta^j$ be the Hartogs expansion of F. Since the disc $\{(z,\zeta) : |\zeta| < 1/e\}$ is included in D, we obtain the coefficient estimates $|f_j(z)| \leq 1/e^j$, $j \geq 0$. It follows that $f_j \in H^{\infty}(\Delta)$, $j \geq 0$. Since the disc $\{(a_k,\zeta) : |\zeta| < 1\}$ is included in D, we obtain the coefficient estimates $|f_j(a_k)| \leq 1$, $k \geq 0$. By (4.5), $|f_j(z)| \leq 1$ for all $z \in \Delta$. The series $\sum f_j \zeta^j$ then converges normally on $\Delta \times \Delta$, and its sum G is an analytic extension of F to $\Delta \times \Delta$. Evaluation of G at any point of $\Delta \times \Delta$ yields a multiplicative functional on $H^{\infty}(D)$, and such functionals have unit norm. It follows that $\|G\|_{\Delta\times\Delta} = \|F\|_D$, so the correspondence $F \to G$ is an isometric isomorphism $H^{\infty}(D) \cong H^{\infty}(\Delta \times \Delta)$.

§5. Compact Sets in C^2, Circled in One Variable

Information on subharmonic functions is rich compared to our understanding of plurisubharmonic functions. If we specialize to the case $n = 1$, we obtain then sharper results concerning the polynomial hull of a compact subset of C^2 circled in one variable.

Theorem 7. Let K, X, R and S be as in Theorem 6, and suppose that $n=1$, so that K is a compact subset of C^2 that is circled in the last variable. Then the set $\{z \in X : R(z) = S(z)\}$ is a closed subset of X which includes the boundary of $\hat{X}$. On each component of $\hat{X} \setminus \{R = S\}$, either $\log S$ is harmonic, or S is identically zero.

Proof. The second conclusion follows immediately from the first, since $-\log S$ is the upper envelope on $\hat{X} \setminus \{R = S\}$ of a Perron family of subharmonic functions. The proof of the first conclusion requires some elementary estimates on harmonic measure. The details are omitted.

Corollary. If K is a compact subset of C^2 that is circled in one variable, then every point of $\hat{K} \setminus K$ lies on an analytic disc in $\hat{K}$.

<u>Proof</u>. If $z_0 \in \hat{X}$ and $|\zeta_0| < e^{-S(z_0)}$, then (z_0,ζ_0) lies on the analytic disc $\{(z_0,\zeta) : |\zeta| < e^{-S(z_0)}\}$. Also, any point of the form $(z_0,0)$, $z_0 \in \hat{X}\setminus X$, lies on an analytic disc of the form $\{(z,0) : |z-z_0| < \varepsilon\}$. All remaining points (z_0,ζ_0) in $R\setminus K$ satisfy $|\zeta_0| = e^{-S(z_0)}$, where either $R(z_0) < S(z_0)$, or $z_0 \notin X$ and $S(z_0) \neq 0$. In either case, Theorem 7 allows us to pass an analytic disc through (z_0,ζ_0) of the form

$$z \to (z,\exp[-\log S(z) - i\,{}^{*}\log S(z)]), \qquad |z-z_0| < \varepsilon,$$

where ${}^{*}\log S$ is an appropriate determination of the conjugate harmonic function of $\log S$.

Q.E.D.

The problem now arises of describing those $F \in C(K)$ which lie in $P(K)$. In connection with this problem, it is appropriate to mention another example of Sibony.

<u>Example</u> (<u>Sibony</u>). There is a compact subset K of C^2, circled in the second variable, such that K is polynomially convex, the interior of K is connected and dense in K, but $A(K) \neq P(K)$.

Here $A(K)$ is the algebra of functions in $C(K)$ which are analytic in the interior of K. The example is constructed as follows. It will be seen from the construction that both K and its interior are topologically trivial.

Let Γ be an arc in the closed unit disc $\bar{\Delta}$ which meets $\partial\Delta$ at only one point, such that there exists a continuous function h on $\bar{\Delta}$ which is analytic on $\Delta\setminus\Gamma$ but which is not analytic on Δ. Let $\{E_j\}_{j=1}^{\infty}$ be a sequence of closed subsets of $\bar{\Delta}$ which increase to $\bar{\Delta}\setminus\Gamma$. By Runge's Theorem, there are polynomials g_j such that $|g_j - j|$ is small on Γ, while $|g_j - 1|$ is small on E_j. Define

$$K = \{(z,\zeta) : |z| \le 1,\ \sup_j |\zeta g_j(z)| \le 1\}.$$

Since K is defined by polynomial inequalities, K is polynomially convex. The interior of K lies over $\Delta\setminus\Gamma$, so that the function h, regarded as a function of z and ζ, lies in $A(K)$. Since h is not analytic on the disc $\Delta\times\{0\} \subset K$, h is not approximable by polynomials. Evidently K is the desired set.

The appropriate question to ask, then, is whether analyticity on every analytic disc in K is sufficient to distinguish which functions in $C(K)$ belong to $P(K)$. We state this problem formally, in terms of a compact subset X of C, the algebra $A = P(X)$, and the entities R, B and Y introduced in Section 2.

(5.1) Under what conditions on X and R is it true that $P(\hat{Y})$ consists of precisely the functions in $C(\hat{Y})$ that are analytic on every analytic disc in $\hat{Y}$?

The following theorem solves the problem in a special case.

Theorem 8. Let X be a compact subset of $\mathbb{C}$ such that X coincides with the boundary of $\hat{X}$. Let $A = P(X)$, and let R and Y be as before. Then $P(\hat{Y})$ consists of precisely the functions in $C(\hat{Y})$ that are analytic on every analytic disc in $\hat{Y}$.

The proof of Theorem 8 proceeds along the following lines. Any $F \in C(\hat{Y})$ which is analytic on each analytic disc in $\hat{Y}$ has a Hartogs expansion $F \sim \sum f_j \zeta^j$. The individual terms $f_j \zeta^j$ of this expansion turn out to be continuous on $\hat{Y}$. By considering uniform approximations to these terms, we find that the problem (5.1) is equivalent, in an appropriate sense, to the following weighted approximation problem in one complex variable.

(5.2) Let g be a complex-valued function defined on $\{z \in \hat{X} : S(z) > 0\}$ such that $g(z)S(z) \to 0$ as $S(z) \to 0$, and such that the restriction of g to each compact set $\{S \geq \varepsilon\}$ is continuous, $\varepsilon > 0$. Is there a sequence of analytic polynomials $\{g_k\}$ such that $g_k S$ converges uniformly to gS?

The proof of Theorem 8 boils down to showing that if $-\log S$ is harmonic on the interior of $\hat{X}$, then the problem (5.2) has an affirmative solution, not only for S but for all S^k, $k > 0$. The affirmative solution of (5.2) in this case depends on Hardy space theory.

§6. Compact Sets in $\mathbb{C}^2$ Invariant under Other Groups

Fix a real number α, and consider the one-parameter group acting on $\mathbb{C}^2$ defined by

$$(z,w) \to (e^{i\alpha\theta}z, e^{i\theta}w), \qquad \theta \in \mathbb{R}. \tag{6.1}$$

We ask for a description of the polynomial hull $\hat{K}$ and the algebra $P(K)$ associated with a compact subset K of $\mathbb{C}^2$ invariant under the transformations (6.1).

If $\alpha = 0$, then being invariant under (6.1) is the same as being circled in the second variable, and the results of the preceding sections give a complete description of $\hat{K}$. If α is irrational, then the transformations in (6.1) are dense in the two-parameter torus group

$$(z,w) \to (e^{i\varphi}z, e^{i\theta}w), \qquad 0 \leq \theta,\ \varphi \leq 2\pi. \tag{6.2}$$

A compact set K invariant under (6.2) is circled, and the description of its polynomial hull is classical. The description of $\hat{K}$ can also be obtained from the case $\alpha = 0$.

Consider next the case in which $\alpha = -1$, so that the group becomes

$$(6.3) \qquad (z,w) \to (e^{-i\theta}z, e^{i\theta}w), \qquad 0 \leq \theta \leq 2\pi.$$

The sets treated by J. Wermer in his lecture are invariant under (6.3). Wermer's techniques can be extended to give a complete description of the polynomial hull of any compact set invariant under (6.3). The description proceeds as follows.

Let K be a compact subset of C^2 that is invariant under the transformation (6.3). Define $F : \hat{K} \to C$ by

$$F(z,w) = zw, \qquad (z,w) \in \hat{K}.$$

Define the fibers M_ζ of $\hat{K}$ by

$$M_\zeta = F^{-1}(\{\zeta\}), \qquad \zeta \in F(\hat{K}).$$

With each fiber, we associate the numbers

$$P(\zeta) = \sup\{|z| : (z,w) \in M_\zeta\}, \qquad \zeta \in F(\hat{K}),$$

$$S(\zeta) = \sup\{|w| : (z,w) \in M_\zeta\}, \qquad \zeta \in F(\hat{K}).$$

It is easy to see that if $\zeta \neq 0$, then M_ζ is the annulus or circle described by

$$M_\zeta = \{(\zeta/w, w) : |\zeta/w| \leq P(\zeta),\ |w| \leq S(\zeta)\}, \qquad \zeta \in F(\hat{K}) \setminus \{0\}.$$

If $\zeta = 0$ happens to belong to $F(\hat{K})$, then M_0 consists of a point, or a disc or two. To describe $\hat{K}$, it suffices then to describe the set $F(\hat{K})$ and the functions P and S. These are determined in terms of the set $F(K)$ and the functions

$$Q(\zeta) = \sup\{|z| : (z,w) \in K,\ zw = \zeta\}, \qquad \zeta \in F(K),$$

$$R(\zeta) = \sup\{|w| : (z,w) \in K,\ zw = \zeta\}, \qquad \zeta \in F(K),$$

by the following theorem.

<u>Theorem</u> 9. Let K be a compact subset of C^2 that is invariant under the transformations (6.3). Define F, P, Q, R and S as above. Then $F(\hat{K})$ coincides with the polynomial hull of $F(K)$. Furthermore, $-\log P$ is the upper envelope of the family of functions u that are subharmonic in a neighborhood of $F(\hat{K})$ and satisfy $u < -\log Q$ on $F(K)$. Similarly, $-\log S$ is the upper envelope of the family of functions v that are subharmonic in a neighborhood of $F(\hat{K})$ and satisfy $v < -\log R$ on $F(K)$.

As before, the set $\{R = S\}$ is a closed subset of $F(K)$ which includes the boundary of $F(\hat{K})$. On each component of $F(\hat{K}) \setminus \{R = S\}$, either S is identically zero, or $\log S$ is harmonic. A similar assertion holds for P and Q, and as before we obtain the following corollary.

Corollary. Every point of $\hat{K} \setminus K$ lies on an analytic disc in $\hat{K}$. Furthermore, $P(K) = C(K)$ if and only if there are no analytic discs in $\hat{K}$.

While the statement of Theorem 9 is analogous to that of Theorem 6, the proof seems to require more than the theory of Hartogs series. It depends on Wermer's idea for filling $\hat{K} \setminus K$ with analytic discs, together with some elementary technical details concerning potential theory in the plane.

There are similar results in the case that K is invariant under (6.1), whenever the parameter α is any negative rational number. However, the polynomial hulls of sets invariant under the group

$$(z,w) \to (e^{i\theta}z, e^{i\theta}w), \qquad 0 \le \theta \le 2\pi,$$

are not yet determined.

BIBLIOGRAPHY

1. H. Bremermann, On a generalized Dirichlet problem for plurisubharmonic functions. Characterization of Šilov boundaries, Trans. A.M.S. 91 (1959), 246-276.

2. H Bremermann, On the conjecture of the equivalence of the plurisubharmonic functions and the Hartogs functions, Math. Ann. 131 (1956), 76-86.

3. D. A. Edwards, Choquet theory for certain spaces of lower semicontinuous functions, in Function Algebras, F. Birtel (ed.), Scott, Foresman and Co., 1965, 300-309.

4. T. W. Gamelin, Uniform algebras spanned by Hartogs series, to appear.

5. T. W. Gamelin, The polynomial hulls of certain subsets of C^2, to appear.

6. F. Hartogs, Zur Theorie der analytischen Funktionen mehrer unabhängiger Veränderlichen, insbesondere über die Darstellung derselben durch Reihen, welche nach Potenzen einer Veränderlichen fortschreiten, Math. Ann. 62 (1906), 1-88.

7. N. Sibony, Prolongement analytique des fonctions holomorphes bornées, C. R. Acad. Sci. Paris 275 (1972), 973-976.

8. N. Sibony, Un exemple de compact polynomialement convexe dans C^2, Bull. Soc. Math. France 103 (1975).

9. V. S. Vladimirov, Methods of the Theory of Functions of Several Complex Variables, M.I.T. Press, Cambridge, Mass., 1966.

10. J. Wermer, Subharmonicity and hulls, Pac. J. Math., to appear.

SOME REMARKS ON INJECTIVE BANACH ALGEBRAS

Sten Kaijser.

Introduction. Injective Banach algebras were introduced by Varopoulos in connection with the study of Q-algebras, i.e. quotients of uniform algebras. Both Q-algebras and injective algebras have been studied intensively during later years and several interesting results have been obtained. We wish to mention in particular the following results:

1. (Cole [5]): Every Q-algebra can be represented as an algebra of operators on a Hilbert space, i.e. as a closed subalgebra of $L(H)$.

2. (Wermer [5]): The Wiener algebra $A(T)$ of functions with absolutely convergent fourier series is not a Q-algebra.

3. (Davie [1]): A Q-algebra is Arens regular and hence bicommutative, i.e. the Arens multiplications in the bidual coincide.
Davie also obtained an intrinsic necessary and sufficient condition for a given Banach algebra to be representable as a Q-algebra [1].

Finally Varopoulos proved that not every closed commutative algebra of operators in Hilbert space is representable as a Q-algebra [4].

The main results concerning injective algebras have all been obtained by Varopoulos who has proved e.g. the following results:

1. Every injective algebra is an operator algebra (and hence Arens regular), i.e. can be represented as a closed subalgebra of $L(H)$.
2. A Banach algebra A is injective $\iff$ for every Banach algebra B, the Tensor product $A \otimes B$ can be given the structure of a normed algebra in the norm induced by the operator norm, or equivalently if the Banach space $A \check{\otimes} B$ is always a Banach algebra.
3. A commutative Banach algebra A is injective $\iff$ there exists a uniform algebra U, an algebra homomorphism $h : U \to A$ and a linear map $L : A \to U$ such that

$h \circ L = 1_A$, i.e. if and only if A is a quotient of a uniform algebra and can also be imbedded as a closed subspace of U.

In this paper we shall study injective algebras. Since some of Varopoulos' papers on injective algebras are rather hard to read we shall begin this paper by giving more abstract and hopefully simpler proofs of some of Varopoulos' results. The main feature of our treatment will be that we shall rely heavily on the results of Grothendiecks "Resume de la theorie metrique des produits tensoriels topologiques". However only the first part of that paper will be used. After that we shall prove some new results concerning injective algebras. Our main theorem is that a 1-injective algebra (see definition (1.2) below) is a uniform algebra even if it is not assumed to be commutative. After that we give some counter-examples to "could be conjectures" inspired by our main theorem. We shall conclude the paper with a study of algebras of differentiable functions. It was proved first by Davie [1] that e.g. the algebra C^1 [0,1] is a Q-algebra, and Varopoulos' observation that C^1 is injective provides a simpler proof of that fact. However the abstract constructions of Davie and Varopoulos give representations of C^1 as the quotient of a very large uniform algebra with e.g. an infinite dimensional Gelfand space, and it is therefore perhaps of some interest that there exists a Banach algebra with a three dimensional Gelfand space and a closed ideal such that the Q-algebra obtained is C^1.

1. <u>Preliminaries and notations</u>. We shall use modified version of the notations of Grothendiecks resume, so for Banach spaces A, B, C etc. with dual space A', B', C' etc. we shall write

$A \hat{\otimes} B$	to denote the projective tensor product of A and B. By definition the dual space of $A \hat{\otimes} B$ is the space
B(A,B)	of bounded bilinear forms on A × B. This space is in turn naturally identified with either one of the spaces
L(A,B')	or
L(B,A')	of bounded linear operators.

Furthermore the space

$A' \hat{\otimes} B$ is often called the space of trace-class operators from A to B. In this connection the space

$A' \hat{\otimes} A$ is particularly important since the bilinear evaluation map on $A' \times A$ has a linear prolongation as the trace. It should be observed that all dualities in spaces of tensor products of Banach spaces depend upon constructing an operator in $A' \hat{\otimes} A$ for some space A. The characterization of $A \hat{\otimes} B$ as the set of those trace-class operators from A' to B that are continuous with respect to the weak *-topology of A' and the weak topology of B is also useful. We shall occasionally also write $A' \check{\hat{\otimes}} B$ to denote the space of bounded operators from A to B. We shall furthermore write

$A \check{\otimes} B$ to denote the injective tensor product of A and B. The space

$A' \check{\otimes} B$ may be defined as the space of those operators from A to B that can be approximated in norm by finite-rank operators and $A \check{\otimes} B$ is likewise the set of those operators from A' to B that are continuous with respect to the weak *-topology of A and the weak topology of B and that can be approximated in norm by finite rank operators. We shall further write

$A' \hat{\hat{\otimes}} B'$ to denote the dual space of $A \check{\otimes} B$. The space $A' \hat{\hat{\otimes}} B'$ is called the space of integral bilinear forms on $A \times B$ and is naturally identified with either one of the spaces

$L^i(A,B')$ of integral operators from A to B' or

$L^i(B,A')$. The space $A \hat{\hat{\otimes}} B$ may be characterized as the set of weak *-continuous integral bilinear forms on $A' \times B'$, where weak *-continuous means that as a function on the product of the unit balls it is continuous. Similarly $A \hat{\hat{\otimes}} B$ may be considered as the space of those integral operators from A' to B that are continuous as functions on the unit ball of A' in its weak *-topology. There is in fact a general principle that whenever '$\overset{\alpha}{\otimes}$' is a tensor norm such that $A' \overset{\alpha}{\otimes} B$ has a characterization as a certain class of operators, then $A \overset{\alpha}{\otimes} B$ may be characterized as those operators from A' to B that lie in $A'' \overset{\alpha}{\otimes} B$ and which are continuous as functions on the unit ball of A' with its weak *-topology.

The space of integral bilinear forms has several useful characterizations. The most important is of course the representations by measures on the product of the dual unit balls which becomes particularly nice as a characterization of integral operators. The simple consequence is that an operator $T : A \to B$ is integral if there exists a probability measure P and a factorization $A \overset{s}{\to} L^{\infty}(dP) \overset{i}{\to} L^{1}(dP) \overset{t}{\to} B$, such that $T = t \circ i \circ s$, where i is the canonical imbedding of L^{∞} into L^{1}. Another useful and important characterization of integral operators arises from the duality with the injective tensor product. For future reference we state the characterization as

Proposition (1.1). Let A and B be Banach spaces and let $T : A \to B$ be a given operator. Then the operator T is integral if there exists a constant K such that for every Banach space C and every operator $S : B \to C$ of finite rank, the operator $S \circ T$ lies in $A' \hat{\otimes} C$ with norm at most $K \cdot \|S\|$. For the converse to hold it suffices that for every operator $S : B \to A$ lying in the closure of the finite rank operators, $S \circ T$ is a trace-class operator.

In terms of the above concepts and notations we shall now define the class of Banach algebras that we are going to study in this paper. First we shall however state the convention that the statement "A is a Banach algebra" means that A is a Banach algebra with a unit. We can now state

Definition (1.2). Let A be a Banach algebra. We shall say that A is K-injective (K a positive real number) if the multiplication π is a continuous linear map of norm at most K, from $A \check{\otimes} A$ to A.

Remark: Since the multiplication in a Banach algebra is bilinear it always has a linear extension to the projective tensor product $A \hat{\otimes} A$. In general this linear map can not be extended to the injective tensor product.

It follows from standard properties of the injective tensor product that a closed subalgebra of a K-injective algebra is K-injective. A Banach algebra is said to be injective if it is K-injective for some K.

2. General properties of injective algebras. We shall start by putting the defining property of an injective algebra in a more useful form as

Proposition (2.1). The Banach algebra A is K-injective if and only if either one of the following equivalent properties hold.

a) For every $S \in A'$ the bilinear form $B_S : A \times A \to \mathbb{C}$, given by $B_S(a,b) = \langle S, ab\rangle$ is integral and of integral norm $\leq K \cdot \|S\|$.
b) For every $S \in A'$ the linear map $L_S : A \to A'$ given by $\langle L_S(a), b\rangle = \langle S, ab\rangle$ is integral and of integral norm $\leq K \cdot \|S\|$.

Proposition (2.1) is an immediate consequence of definition (1.2) and the duality between injective tensor products and integral bilinear forms, and of the interpretation of bilinear forms as linear operators. The idea of studying a Banach algebra in terms of the family of bilinear forms B_S has been extensively used by Varopoulos in the study of Q-algebras, injective algebras and operator algebras. As another application of this method we shall prove the following

Proposition (2.2). Let A be a Banach algebra. Then A is Arens regular, i.e. the two Arens multiplications on A'' coincide, if and only if for every $S \in A'$, the bilinear form B_S is weakly compact or equivalently, if for every $S \in A'$, the linear map $L_S : A \to A'$ is weakly compact.

For the proof of proposition (2.2) we first remark that given any bilinear form B on a pair of Banach spaces X, Y, there exists two canonical prolongations B_r and B_l of B to the pair X'', Y'' and that the Arens regularity of A means exactly that the two prolongations coincide. To prove the proposition it therefore suffices to apply the following lemma which is stated between parentheses in [2]:

Lemma (2.3). Let X and Y be Banach spaces and let B be a bilinear form on $X \times Y$. Then the canonical prolongations B_r and B_l coincide on $X'' \times Y''$ if and only if B is weakly compact.

Proof: By definition B is said to be weakly compact if the linear map $L_B : X \to Y'$ is weakly compact or equivalently if the linear map $R_B : Y \to X'$ is weakly compact. We define now

$$B_l(x'',y'') = \langle L_B'' x'', y'' \rangle$$

and

$$B_r(x'',y'') = \langle x'', R_B'' y'' \rangle .$$

From the definitons it follows then that B_l is a separately continuous function on $X'' \times Y''$ for the weak $*$-topology of X'' and the weak topology of Y'', while B_r is separately continuous for the weak topology on X'' and the weak $*$-topology on Y''. If the functions are to coincide they will then have to be separately continuous for the weak $*$-topologies, and this means that e.g. L_B'' maps X'' into Y' and this is equivalent to the weak compactness of L_B.

Remark: When applying proposition (2.1) it suffices to have condition a) or b) holding for all S in a norm-determining subset of A'. When applying proposition (2.2) it is usually necessary to check the conditions for e.g. a norm-dense subset of the unit ball of A' or the extreme points of the unit ball of A'.

As a first application of proposition (2.1) we can give a simple proof of the following result of Varopoulos [4]:

Proposition (2.4): Every injective algebra can be represented as a closed subalgebra of $L(H)$ for some Hilbert space H.

Proof: Let A be K-injective. We shall find for every $a \in A$ a Hilbert space H_a and a representation T_A of A on H_a such that $\|T_A\| \geq K^{-1}\|a\|$. Applying the Hahn-Banach theorem we may choose S in the unit ball of A' such that $\langle S, a\rangle = \|a\|$. Since A is injective the operator $L_S : A \to A'$ is integral and thus has a factorization

$$A \xrightarrow{s} L^\infty(dP) \xrightarrow{i} L^1(dP) \xrightarrow{t} A'$$

Between L^∞ and L^1 we have L^2 which gives us a factorization

$$A \xrightarrow{u} L^2(dP) \xrightarrow{v} A' ,$$

such that $v \circ u = L_S$ and $\|u\| \cdot \|v\| \leq K$. Projecting onto $\mathrm{Ker}(v)^\perp$

we get a factorization

$$A \overset{u'}{\rightarrow} H \overset{v'}{\rightarrow} A' \quad ,$$

such that $v' \circ u' = L_S$, $\|u'\| = 1$ and $\|v'\| \leq K$, and such that v' is an injective linear map. It is now easy to see that u' gives a representation of A on H, that $u'(1)$ is a cyclic vector in H, and that $(T_a(u'(1))|u'(1)) = \langle S,a \rangle = \|a\|$ and hence $\|T_a\| \geq K^{-1}\|a\|$, and this proves the theorem. Q.

As another application of proposition (2.1) we shall also give a simplified (i.e. less technical) proof of Varopoulos' beautiful characterization of injective algebras.

<u>Proposition (2.5)</u>: Let A be a Banach algebra. Then A is injective $\Leftrightarrow$ For every Banach algebra B the injective tensor product $A \check{\otimes} B$ is a Banach algebra, under the multiplication induced from the multiplications in A and B.

<u>Proof</u>: $\Rightarrow$) Since the multiplication will be associative if it exists, it suffices to prove the existense of a bilinear map

$$\pi : A \check{\otimes} B \times A \check{\otimes} B \rightarrow A \check{\otimes} B$$

such that $\pi(a \otimes b, a' \otimes b') = aa' \otimes bb'$. The existence of such a map follows from the following simple

<u>Lemma (2.6)</u>: Let A, B, C and D be Banach spaces. Then there exists a natural linear map $T : (A \check{\otimes} B) \hat{\otimes} (C \check{\otimes} D) \rightarrow (A \check{\otimes} C) \check{\otimes} (B \hat{\otimes} D)$.

<u>Proof</u>: Let $S \in A \check{\otimes} B$ and $T \in C \check{\otimes} D$. Since elements of finite rank are norm dense in these spaces, we may assume S and T to be of finite rank and we then consider e.g. S as a weak $*$-continuous operator from A' to C. We consider then $S \otimes T$ as a weak $*$-continuous operator from $A' \hat{\otimes} C'$ to $B \hat{\otimes} D$. As such it lies in $(A' \hat{\otimes} C')' \check{\otimes} (B \hat{\otimes} D)$, being of finite rank it then lies in $(A'' \check{\otimes} C'') \check{\otimes} (B \hat{\otimes} D)$ and from the weak $*$-continuity it follows that it lies in $(A \check{\otimes} C) \check{\otimes} (B \hat{\otimes} D)$, and this proves the lemma. Q.

For the proof of the proposition we now define

$$\pi: (A \overset{\vee}{\otimes} B) \overset{\wedge}{\otimes} (A \overset{\vee}{\otimes} B) \to A \overset{\vee}{\otimes} B, \qquad \text{by the factorization}$$

$$(A \overset{\vee}{\otimes} B) \overset{\wedge}{\otimes} (A \overset{\vee}{\otimes} B) \to (A \overset{\vee}{\otimes} A) \overset{\vee}{\otimes} (B \overset{\wedge}{\otimes} B) \xrightarrow{\pi_A \otimes \pi_B} A \overset{\vee}{\otimes} B.$$

<=) For the converse we shall use proposition (1.1) in form of the following

Lemma (2.7): Let A and B be Banach spaces and let β be a bilinear form on $A \times B$. Then β is integral and of integral norm $\leq K$ if for any pair C, D of Banach spaces, and any bilinear form γ on $C \times D$, the multilinear form $\beta \otimes \gamma$ on $A \times B \times C \times D$ has a continuous extension of norm $\leq K \cdot \|\gamma\|$ to the space $(A \overset{\vee}{\otimes} C) \overset{\wedge}{\otimes} (B \overset{\vee}{\otimes} D)$, and it suffices that this assumption holds for any bilinear form on $G \times G$ where G is a certain universal Banach space.

Proof: We consider β as a linear map from A to B', γ as a linear map from D to C', and we shall prove that β is an integral map. We let $S \in (A \overset{\vee}{\otimes} C)$, $T \in (B \overset{\vee}{\otimes} D)$ be finite rank operators. Considering S as an operator from C to A', and T as an operator from B' to D, it is easy to see that the form $\beta \otimes \gamma$ can be obtained as the trace of the following composed operator

$$A \overset{\beta}{\to} B' \overset{T}{\to} D \overset{\gamma}{\to} C' \overset{S}{\to} A .$$

By assumption we have $|\mathrm{Tr}(S \circ \gamma \circ T \circ \beta)| \leq K \cdot \|S\| \cdot \|\gamma\| \cdot \|T\|$, and if we could apply proposition (1.1), the lemma would be proved. What remains to be proved is that there exist spaces C and D, such that every finite rank operator $F : B' \to A$ has a factorization over D and C' in such a way that $\|S\| \cdot \|\gamma\| \cdot \|T\| \leq (1+\varepsilon)\|F\|$.

To do this we shall use a certain universal Banach space G with certain nice (i.e. bad) properties. The existence of the Banach space G has been observed by several authors e.g. Grothendieck [2] or Figiel [3], where a construction is given. The Banach space G has the following properties:

a) G is separable
b) There exists an isometric isomorphism $U : G \to G'$ such that $U' = U$, so in particular G is reflexive.

c) Given Banach spaces X and Y and finite rank operator $F : X \to Y$ such that $\|F\| < 1$, there exist operators R, S, T all of finite rank, all contractions such that

$$X \overset{R}{\to} G \overset{S}{\to} G \overset{T}{\to} Y \text{ , and } T \circ S \circ R = F$$

(Essentially G is an l^2-sum of all finite-dimensional spaces, or rather a dense subset of the set of all finite dimensional Banach spaces). To finish the proof of proposition (2.5) it suffices according to the proof of lemma (2.9), to prove that for any bilinear form γ on $G \times G$, the quadrilinear form $B_S \otimes \gamma$ has norm $\leq K \cdot \|S\| \cdot \|\gamma\|$ on the space $(A \check{\otimes} G) \hat{\otimes} (A \check{\otimes} G)$, B_S being the bilinear form defined in propositon (2.1). To do this we simply imbed G in its tensor algebra $T(G)$ and use the fact (observed by Varopoulos) that every bilinear form on G can be represented as B_T (with the notations of propositon (2.1)) for some $T \in T(G)'$. The sought conclusion follows then from the assumption that $A \check{\otimes} T(G)$ is a Banach algebra. Q.

Remark: The above proof is only a slight modification of the proof given by Varopoulos, but it also gives slightly more since it follows that A is injective if $A \check{\otimes} T(G)$ is a Banach algebra.

3. <u>1-injective Banach algebra.</u> In the preceding section we have used the term Banach algebra to denote a Banach space with a continuous associative product. From now on we shall however use the term Banach algebra to denote a Banach space with an associative product of norm 1, and having a unit. This will in particular permit us to use the notion of a state, which is simply an element $f \in A'$, such that $f(1) = \|f\| = 1$. It is clear that the set of all states is a weak $*$-compact convex subset of the unit ball of A'. We can now state and prove the following

Theorem (3.1): A 1-injective Banach algebra is a uniform algebra.

Remark 1: It follows that a 1-injective algebra is commutative and semisimple.

Remark 2: The algebra need not even be associative.

Remark 3: Assuming commutativity the theorem would be almost obvious from Varopoulos characterization as a complemented quotient of a uniform algebra.

Proof: Let A be a 1-injective Banach algebra with dual space A', and with state-space (set of all states) S. We begin by proving that for every $a \in A$, there exists a state $s \in S$, such that $|\langle s,a\rangle| = \|a\|$. To do this we choose first $f \in A'$ such that $\|f\| = 1$, and $\langle f,a\rangle = \|a\|$. The bilinear form B_f on $A \times A$ (where as before $B_f(a,b) = \langle f,ab\rangle$) is then integral, and of integral norm 1. This implies then that there exists a probability measure P on $A'_1 \times A'_1$ (where A'_1 is the closed unit ball of A'), such that

$$B_f(x,y) = \int \langle a',x\rangle \langle b',y\rangle \, dP(a',b').$$

Choosing now first $x = 1$, $y = a$, it follows that on the support of P we have $|\langle a',1\rangle| = 1$, $|\langle b',a\rangle| = \|a\|$, while choosing $y = 1$, $x = a$ we have $|\langle a',a\rangle| = \|a\|$, $|\langle b',1\rangle| = 1$, and dropping the assumption that P is positive we may assume P to be supported by $S \times S$, and this implies that for some $s \in S$ we have $|\langle s,a\rangle| = \|a\|$.
Let now e be an extreme point of S. We shall prove that e is a multiplicative linear functional. Towards this we choose an integral representation of the bilinear form B_e as

$$\int s_1 \otimes s_2 dP(s_1,s_2),$$

where s_1 and s_2 are states as we see from the assumption $B_e(1,1) = 1$. Since furthermore $\langle e,x\rangle = B_e(1,x)$ it follows from the assumption that e is an extreme point of S, that P is supported by the set $S \times \{e\}$, and since we also have $\langle e,x\rangle = B_e(x,1)$ we have finally that P is supported by $\{e\} \times \{e\}$, so that $B_e(x,y) = \langle e,x\rangle \langle e,y\rangle$, and this implies that e is multiplicative.

We have thus proved that the norm of every element is given by a multiplicative linear functional so we have proved that A is a uniform algebra. Q.

Remark 1: The theorem would not be true if we admitted a Banach algebra without unit to be 1-injective as is seen e.g. by considering the Banach space ℓ^1 as a Banach algebra under pointwise multiplication.

Remark 2: It would be interesting to know to what extent the number 1 is critical for the conclusion of the theorem, or if one can increase the injectivity constant slightly and still have e.g. commutivity or semisimplicity. From 2- and 3-dimensional examples it follows that there exists a non semi-simple K-injective algebra with $K \leq 2/\sqrt{3}$, and a non-commutative K-injective algebra with $K \leq \sqrt{2}$.

4. Banach algebras of differentiable functions. Let M be a compact manifold (with or without boundary) and let A be the Banach algebra $C^k(M)$ of k times continuously differentiable functions on M. Davie [1] proved that A is a Q-algebra, and Varopoulos improved this result by proving that A is in fact injective. The representations of A as a quotient of a uniform algebra U_A, are however rather complicated. In particular the Gelfand space of U_A will contain at least the unit ball of A'. We shall here prove that A can also be represented as a quotient of a rather small uniform algebra with a Gelfand space consisting of $M \times D^n$, where M is the given manifold of dimension n, and D^n is the polydisc of dimension n. Since the algebra $C^1(\Pi)$ ($= C^1[0,1]$) is typical, we shall only consider that algebra for a while. We start then by considering the algebra $A(\Pi \times D)$ consisting of all functions on $\Pi \times D$ that are continuous and are analytic on D^0 for each fixed $t \in \Pi$. The elements of A ($= A(\Pi \times D)$) have a representation by a power series

$$f(t,z) = \sum_{k=0}^{\infty} f_k(t) z^k .$$

Let I_2 be the ideal of all functions in A such that $f_0(t) = f_1(t) = 0$ for all t. Then A/I_2 is the Banach algebra C_2 of all pairs of continuous functions on Π, with the multiplication $(f_0,f_1) \cdot (g_0,g_1) = (f_0 g_0, f_0 f_1 + f_1 g_0)$. Putting $\|(f_0,f_1)\| = \max(|f_0(t)| + |f_1(t)|)$ we clearly have a Banach algebra, and defining ${}^{\Pi}L : C_2 \to A$ by

$$L(f_0,f_1)(t,z) = f_0(t) + z f_1(t)$$

we have a linear right inverse of $h : A \to C_2$. C_2 is therefore injective. Now $C^1(\Pi)$ is easily seen to be a closed subalgebra of C_2, and it follows that starting instead from the algebra $A_1(\Pi \times D)$

of all continuous functions on $\Pi \times D$ that are analytic in D^{0} for each $t \in \Pi$, and that have a power series representation

$$f(t,z) = \sum_{k=0}^{\infty} f_k(t) z^k$$

such that $f_0(t) = f_0(0) + \int_0^t f_1(u)du$, we see that $A_1/I_2 = C^1(\Pi)$.

Essentially the same representation also holds for any $C^k(M)$. However, it does not seem possible to represent Lipschitz algebras by the same method.

We shall conclude this paper by remarking that even though the map L above does have norm 1, this does not imply that C^1 is 1-injective. To see this it suffices to observe that the usual norm for C^1 (and also the slightly smaller norm defined above for C_2) is not the quotient norm from A_1. The quotient norm can however be computed and turns out to be given by

$$||f|| = \max_{t\in\Pi}\left(\left(|f(t)|^2 + \frac{|f'(t)|^2}{4}\right)^{1/2} + \frac{|f'(t)|}{2}\right) .$$

With this norm C^1 turns out to be $\frac{5}{4}$-injective.

REFERENCES.

[1] A.M. Davie, Quotient algebras of uniform algebras, J. London Math. Soc. 7 (1), 1973, pp 31 - 40.

[2] A. Grothendieck, Resumé de la théorie metrique des produits tensorielles topologiques, Bol. Soc. Mat. Sao Paolo 8 (1956), pp 1 - 79.

[3] T. Figiel, Factorization of compact operators, Stud. Math. 45 (1973) pp 191 - 210.

[4] N. Varopoulos, C.R.A.S. 274, 275, 276 (1972, 73).

[5] J. Wermer, Quotient algebras of uniform algebras, symp. on function algebras and rational approximation, University of Michigan 1969.

Some remarks on automatic continuity.

K.B. Laursen, Institute of Mathematics,
University of Copenhagen.

Introduction.

Much of the recent work on automatic continuity has centered around the problem of continuity/discontinuity of linear operators intertwining with certain continuous linear operators.

This is a natural development of the field which traditionally spends most of its energy on problems related to the continuity of algebra homomorphisms (which intertwine with the algebra multiplication).

At the same time these results may be viewed as statements about linear operators with certain invariant subspaces. Perhaps the most general result along these lines may be found in Allan Sinclair's survey of this area [3]; we shall return to his continuity result later.

Here we present the beginning of an attempt to investigate the extent to which assumptions concerning invariant subspaces may be suppressed. At this point of the development much of what will be said amounts to a reorganization of known facts. Certainly, anyone familiar with the early versions of [3] will recognize the debt owed this work.

The basic result here is the stability theorem (2.1), which is put to use through its corollary, Proposition 3.2. The scope of 3.2 is not at all clear yet; but at least it is strong enough to yield a simple proof of Sinclair's basic continuity result alluded to above (cf. 3.6) and to allow some simplification of Sinclair's proof [3] of Johnson's result [1] on continuity of irreducible representations (cf. 4.3).

1. The separating space.

We state here, without proofs, the basic facts concerning the separating space of a linear map.

If $S: X \to Y$ is a linear map from the Banach space X to the Banach space Y then the separating space of S is

$$\mathfrak{S}(S) = \{y \in Y \mid \exists \{x_n\} \subset X, x_n \to 0 \text{ and } Sx_n \to y\}$$

It is easy to see that $\mathfrak{S}$ is a closed linear subspace of Y. The closed graph theorem tells us that $\mathfrak{S}(S) = \{0\}$ if and only if S is continuous.

If X_1, Y_1 are Banach spaces and $T: X_1 \to X$, $R: Y \to Y_1$ are bounded linear operators then $\mathfrak{S}(ST) \subseteq \mathfrak{S}(S)$ and $[R\mathfrak{S}(S)]^- = \mathfrak{S}(RS)$ (where $^-$ denotes closure).

From the last fact it follows that if $Y_0 \subseteq Y$ is a closed linear subspace and $Q: Y \to Y/Y_0$ is the canonical map, then $QS: X \to Y/Y_0$ is continuous if and only if $\mathfrak{S}(S) \subseteq Y_0$.

Finally we shall find use for the following result of Patterson. A proof may be found in [3].

Proposition. Let $Z_1, \cdots, Z_n$ be Banach spaces and $T_1, \cdots, T_n$ be bounded linear operators from $Z_1, \cdots, Z_n$ into X such that

$$X = T_1 Z_1 + \cdots + T_n Z_n.$$

Let $S: X \to Y$ be linear. Then

$$\mathfrak{S}(S) = [\mathfrak{S}(ST_1) + \cdots + \mathfrak{S}(ST_n)]^-.$$

2. The stability theorem.

The following result has been basic to much of the most recent work in the field, e.g. Sinclair's work on homomorphisms from $C_o(\mathbb{R})$ [2] (where it is not stated explicitly).

Let $S: X \to Y$ be a linear mapping between the Banach spaces X and Y and let $\{X_i\}_{i=1}^{\infty}$ be a sequence of Banach spaces with $X_o = X$. Suppose we have bounded linear operators $T_i: X_i \to X_{i-1}$, $i = 1,2,\cdots$. If $\mathcal{S}_n = \mathcal{S}(ST_1\cdots T_n)$, $n = 1,2,\cdots$ and $\mathcal{S} = \mathcal{S}(S)$ then clearly

$$\mathcal{S} \supseteq \mathcal{S}_1 \supseteq\cdots\supseteq \mathcal{S}_n \supseteq\cdots$$

But in fact, much more is true.

Proposition 2.1 (Stability theorem).

$$\exists N: n \geq N \Rightarrow \mathcal{S}_n = \mathcal{S}_{n+1}.$$

Proof: If no such N may be found we may assume (by bunching the T_n's, if necessary) that

$$\mathcal{S} \supsetneq \mathcal{S}_2 \supsetneq\cdots\supsetneq \mathcal{S}_n \supsetneq\cdots$$

Letting $Q_n: Y \to Y/\mathcal{S}_n$ denote the natural quotient maps we note that $Q_nST_1\cdots T_m$ is discontinuous for $m < n$, while $Q_nST_1\cdots T_m$ is continuous for $m \geq n$ (section 1).

Assuming, as we may, that $\|T_n\| = 1$ for all n we select inductively y_n satisfying

i) $\|y_n\| < 2^{-n}$

ii) $$\|Q_n ST_1 \cdots T_{n-1} y_n\| \geq n + \|Q_n ST_1 \cdots T_n\| + \sum_{p=2}^{n-1} \|Q_n ST_1 \cdots T_{p-1} y_p\|$$

for $n = 2,3,\cdots$.

Then let $x_o = \sum_{n=2}^{\infty} T \cdots T_{n-1} y_n$ and note that

$$\|Sx_o\| \geq \|Q_n Sx_o\| = \| Q_n S \sum_{m=2}^{n-1} T_1 \cdots T_{m-1} y_m + Q_n ST_1 \cdots T_{n-1} y_n + Q_n S \sum_{m>n} T_1 \cdots T_{m-1} y_m \|$$

$$\geq n + \| Q_n ST_1 \cdots T_n \| + \sum_{p=2}^{n-1} \| Q_n ST_1 \cdots T_{p-1} y_p \| - \sum_{m=2}^{n-1} \| Q_n ST_1 \cdots T_{m-1} y_m \| - \| Q_n ST_1 \cdots T_n \| \, \| \sum_m T_{n+1} \cdots T_m y_{m+1} \|$$

$$\geq n.$$

3. Singularity points.

The following condition is an abstraction of the relationship between a closed ideal in a commutative regular Banach algebra and its hull (the set of maximal modular ideals containing it, i.e. the 'zero-set' of the ideal).

Condition 3.1 Suppose X is a Banach space, Ω is a regular Hausdorff topological space and Γ is a basis for the topology on Ω. Suppose $F \to X(F)$ is a mapping from Γ into the closed linear subspaces of X satisfying:

a) $F_1, F_2 \in \Gamma: F_1 \subseteq F_2 => X(F_2) \subseteq X(F_1)$.

b) whenever $F_1, \cdots, F_n$ in Γ satisfy $F_i^- \cap F_j^- = \emptyset$ for all $i \neq j$ (where $^-$ denotes closure) then

$$X(F_1) \cap \cdots \cap X(F_{n-1}) + X(F_n) = X.$$

If we combine this condition with the stability theorem the following stability result emerges.

Proposition 3.2 Let X and Ω satisfy condition 3.1 and let $S: X \to Y$ be a linear mapping into the Banach space Y. Suppose $\{F_i\}$ is an infinite sequence of sets from Γ which have disjoint closures. Consider the restrictions of S to the subspaces $X(F_n)$. Then

$$\exists N: n \geq N \to \mathfrak{S}(S) = \mathfrak{S}(S|_{X(F_n)}).$$

Proof: For each $n = 2,3,\cdots$ we have

$$X = X(F_1) \cap \cdots \cap X(F_{n-1}) + X(F_n) = X_n + X(F_n).$$

If $I_n : X_n \to X$ and $J_n : X(F_n) \to X$ denote the injections we may view I_n as a mapping from X_n into X_{n-1} so that by the stability theorem (Proposition 2.1) there is a N so that $\mathfrak{S}(S|_{X_n}) = \mathfrak{S}(S|_{X_{n+1}})$ for all $n \geq N$.

By Patterson's result (Proposition, section 1)

$$\mathfrak{S}(S) = (\mathfrak{S}(S|_{X_n}) + \mathfrak{S}(S|_{X(F_n)}))^-;$$

moreover, since $X_n \subseteq X(F_{n-1})$ it is clear that $\mathfrak{S}(S|_{X_n}) \subset \mathfrak{S}(S|_{X(F_{n-1})})$. Combining these fact, we obtain

$$\mathfrak{S}(S) = \mathfrak{S}(S|_{X(F_n)})$$

whenever $n \geq N$.

This proposition tells us that under appropriate assumptions the restrictions of S to certain sequences of subspaces will eventually capture the entire separating space.

For lack of a better term we shall call a set $F \in \Gamma$ a good set if its corresponding subspace $X(F)$ has this capturing property. Specifically:

Definition 3.3 $F \in \Gamma$ is good if $(S|_{X(F)}) = \mathfrak{S}(S)$. Also, a point $w \in \Omega$ will be called a singularity point of S if w lies in no good set.

In terms of singularity points, Proposition 3.2 says the following:

Corollary 3.4 S has at most finitely many singularity points.

Sketch of proof: If there are infinitely many singularity points a standard separation argument will produce a sequence $\{F_n\}$ of sets in Γ with disjoint closures each with the property that $\mathfrak{S}(S|_{X(F_n)}) \neq \mathfrak{S}(S)$, in contradiction with Proposition 3.2.

Corollary 3.4 is a variant of Sinclair's basic continuity result [3, Theorem 2.3]. His setting is the following:

Condition 3.5 Suppose X,Y are Banach spaces, Ω a regular Hausdorff space, S a linear map from X to Y. Suppose X and Ω are related as in Condition 3.1; suppose, moreover, that there is a mapping from Γ into the closed subspaces of Y with respect to which S satisfies the invariance condition that $SX(F) \subseteq Y(F)$ for every $F \in \Gamma$. Finally, $F \in \Gamma$ contains no discontinuity points for S if $\mathfrak{S}(S) \subseteq Y(F)$.

Actually, Sinclair's continuity result follows from Corollary 3.4.

Corollary 3.6 (Sinclair) If S,X,Y,Ω satisfy Condition 3.5, then S has finitely many discontinuity points.

Proof: If not, then an infinite sequence of discontinuity points may be separated by sets F_n with disjoints closures. For all large n, $\mathfrak{S}(S) = \mathfrak{S}(S|_{X(F_n)})$, by Corollary 3.4. But clearly $\mathfrak{S}(S|_{X(F_n)}) \subseteq (SX(F_n))^-$ and since $Y(F_n)$ is closed and contains $S(X(F_n))$ we get $\mathfrak{S}(S) \subseteq Y(F_n)$ for all large n. This contradiction establishes the claim.

Sinclair has put 3.6 to many good uses, among them a proof of Barry Johnson's famous result on the automatic continuity of irreducible representations of Banach algebra, leading to a solution of the uniqueness of norm problem for semi-simple Banach algebras. In the rest of this note we present a slightly different approach, based directly on 3.4.

4. Continuity results.

If we impose one additional set of assumptions then we can prove certain continuity results which might tend to justify the use of the terms 'singularity points' and 'good sets'.

Condition 4.1 Suppose X and Ω are related as in 3.1. Suppose in addition that S satisfies:

$$\text{whenever } \bigcup_\alpha F_\alpha = \Omega \text{ with } F_\alpha \in \Gamma \text{ then}$$

$$\bigcap_\alpha (SX(F_\alpha))^- = \{0\}.$$

Remark: If Ω is compact it is enough to consider finite collections $\{F_\alpha\}$.

We then have an analogue of [3, Corollary 2.5].

Proposition 4.2. Let $F = \{\lambda_1, \cdots, \lambda_p\}$ be the set of singularity points of S and suppose 4.1 is satisfied. if $F \subseteq \bigcup_{j=1}^m W_j$, then $S|_{X(W_1) \cap \cdots \cap X(W_m)}$ is continuous.

Proof: For notational simplicity suppose $j = p$ and $\lambda_k \in W_k$, $k = 1, \cdots, p$. For each k choose $U_k \subseteq U_k^- \subseteq W_k$ such that $\lambda_k \in U_k \in \Gamma$. For each $w \in \Omega \setminus \bigcup_{k=1}^p U_k$ we can find a good set $E_w \in \Gamma$. Since $\Omega = \bigcup E_w \cup \bigcup_{k=1}^p U_k$ we get that

$$\{0\} = \bigcap_w (SX(E_w))^- \cap \bigcap_{k=1}^p (SX(U_k))^-.$$

Since $\mathcal{S}(S) \subseteq S(X(E_w))^-$ for each w we get

$$\mathcal{S}(S) \cap \bigcap_{k=1}^{p} S(X(U_k))^- = \{0\}$$

and hence

$$\mathcal{S}(S|_{X(W_1)\cap\cdots\cap X(W_p)}) \subseteq S(\bigcap_{k=1}^{p} X(W_k))^- \cap \mathcal{S}(S)$$

$$\subseteq S(\bigcap_{k=1}^{p} X(U_k))^- \cap \mathcal{S}(S) = \{0\}$$

from which the result follows, by section 1.

Finally we present, in outline, a proof of Johnson's result on continuity of irreducible representations of Banach algebras, based on 4.2.

Proposition 4.3 (Johnson [1]) Let A be a Banach algebra, X a normed linear space and $\Pi: A \to L(X)$ an irreducible representation ($L(X)$ denotes the bounded linear operators on X). Then Π is continuous.

Sketch of proof: If X is finite dimensional the result is standard, so we assume X to be infinite dimensional. Also, by the uniform boundedness principle it suffices to show that $a \to \Pi(a)x$ is continuous for each $x \in X$. To show the continuity of these maps, it is enough to do it for one $x \in X$; the result for any $x \in X$ then follows by the irreducibility of Π.

Let $\{x_n\}$ be an infinite sequence of linearly independent vectors in X, let $Z = \text{span}(x_i)$ (algebraic span) and let $Y = L(Z,X)$. Define $S: A \to Y$ by $S(a)(z) = \Pi(a)z$ for each $z \in Z$.

We take Ω to be $\{1,2,\cdots\}$ with the discrete topology and for each finite $F \subset \Omega$ we let

$$A(F) = \{a \in A \mid \pi(a)x_j = 0 \quad \text{for all} \quad j \in F\}.$$

Since $A(\{j\})$ is a maximal modular left ideal for any j, it is easy to check that Condition 3.1 is satisfied. Similarly, Condition 4.1 may be easily checked. If F_0 denotes the singularity set of S and if $k \notin F_0$ then by Proposition 4.2 the mapping

$$a \to \pi(a)x_k$$

is continuous on the set $A(F_0) = \{a \in A \mid \pi(a)x_j = 0$ for all $j \in F\}$. By irreducibility we may find $b \in A$ such that

$$\pi(b)x_j = x_j \qquad j \in F$$

$$\pi(b)x_k = 0$$

Clearly, $a - ab \in A(F_0)$ so

$$\| \pi(a)x_k \| = \| \pi(a-ab)x_k \| \leq C\| a-ab \| \, \| x_k \|$$

$$\leq C\| a \| (1+\| b \|)\| x_k \|$$

which shows that $a \to \pi(a)x_k$ is a continuous map.

5. Some open problems.

1. Under what conditions on S is the collection of good sets closed under unions?
   Will a condition like $S(X(E_j) \cap X(E_i))^- = S(X(E_j))^- \cap S(X(E_i))^-$ suffice?

2. When does 4.1 have a chance of being true (assuming Ω non-discrete)?

3. (Sinclair) In connection with the stability theorem (2.1), is it possible to use the assumption ST_1 continuous to conclude that

$$\{\| ST_1 \cdots T_n \| / \| T_1 \cdots T_n \| \}$$

is a bounded set?

References.

[1] B.E. Johnson : Uniqueness of the (complete) norm topology, Bull. Amer. Math. Soc. 73 (1967), 537-539.

[2] A.M. Sinclair : Homomorphisms of $C_0(\mathbb{R})$, Proc. London Math.Soc. (to appear)

[3] A.M. Sinclair : Notes on automatic continuity (to appear in the London Math.Soc. lecture notes series).

ON BANACH SPACE PROPERTIES OF UNIFORM ALGEBRAS

A. Pełczyński (Warszawa)

The main result of the present paper is Theorem 1 below which generalizes both the recent result of the author [9] for the disc algebra and the recent result of Kisliakov [6], who proved that if A is a uniform algebra on a compact Hausdorff space and A has a Gleason part with at least two points, then A in this particular situation is uncomplemented in $C(X)$. It supports the following

<u>Conjecture.</u> If A is a uniform algebra on X which as a Banach space is isomorphic to a quotient of a $C(K)$ space for some compact Hausdorff space K, then $A = C(X)$.

We are able to prove

<u>Theorem 1.</u> Let A be a uniform algebra on a compact Hausdorff space X and assume that there exists a Gleason part of A which contains at least two different points. Then neither

(a) A is isomorphic as a Banach space to a quotient of a $C(K)$-space,

nor

(b) A regarded as a real Banach space is isomorphic to a complemented subspace of a Banach lattice,

nor

(c) A admits a Gordon-Lewis unconditional structure, i.e. there exists a constant $k > 0$ such that for every finite dimensional subspace F of A there exist a finite dimensional space B_F with a basis (b_j) and operators $S_F : F \to B_F$ and $T_F : B_F \to A$ such that

$$T_F S_F(f) = f \quad \text{for} \quad f \in F \; ; \; \|T_F\| \, \|S_F\| \leq k$$

and for every sequence of scalars (c_j)

$$\|\Sigma\, c_j b_j\| = \|\Sigma |c_j| b_j\| .$$

In the present paper "operator" always stands for bounded linear operator.

The proof of Theorem 1 is based upon the following

Criterion. (Gordon-Lewis [5], Figiel-Johnson-Tzafriri [3]).
Let A be a Banach space which has one of the properties (a), (b), (c) stated in Theorem 1. Then every absolutely summing operator from A into an arbitrary Banach space H factors locally through L^1-spaces.
Recall that an operator $u : A \to H$ is said to be absolutely summing if there exists a positive Borel measure ρ on a compact Hausdorff space X such that A can be identified with a closed linear subspace of $C(X)$ (precisely A is isometrically isomorphic to a subspace of $C(X)$) and

$$\int_X |f(x)|d\rho \geq \|u(f)\| \quad \text{for every } f \in A.$$

Next recall that an operator $u : A \to H$ locally factors through L^1-spaces if there exists a constant $k > 0$ such that for every finite dimensional subspace F of A there exist operators $S_F : F \to l^1_{N(F)}$ and $T_F : l^1_{N(F)} \to H$ such that $T_F S_F(f) = u(f)$ for $f \in F$, $\|S_F\|\,\|T_F\| \leq k$ and $N(F) < \infty$ (here l^1_k denotes the k-dimensional vector space with the norm $\|(c_1,c_2,\ldots,c_k)\| = \sum_j |c_j|$).

The standard ultrafilters technique yields (cf. [3], [5])

Lemma 1. If H is a dual Banach space and if $u : A \to H$ is an operator which locally factors through L^1-spaces then u factors through an L^1-space, i.e. there exists a space $L^1(\nu) = L^1(\nu,X,\Sigma)$ (ν not necessarily sigma finite) and operators $S : A \to L^1(\nu)$ and $T : L^1(\nu) \to H$ such that $TS = u$.
We shall also use the following trivial fact

Lemma 2. Let $u : A \to H$ be an operator which locally factors through L^1-spaces.
Let $U_n : \tilde{A} \to A$ and $V_n : H \to \tilde{H}$ ($A,H,\tilde{A},\tilde{H}$ - arbitrary Banach spaces) be operators such that for every $f \in \tilde{A}$ there exists the strong limit $K(f) = \lim_n K_n(f)$ where $K_n = V_n u U_n$ for $n = 1, 2, \ldots$ and $\sup_n \|U_n\|\,\|V_n\| < \infty$. Then the operator $K : \tilde{A} \to \tilde{H}$ factors locally through L^1-spaces.

In the sequel we shall denote by $A(D)$ the disc algebra i.e. the uniform closure of polynomials on the unit disc $D = \{z \in \mathbb{C} : |z| \le 1\}$. We shall regard $A(D)$ as the uniform algebra on ∂D identifying each $f \in A(D)$ with its restriction to the unit circle $\partial D = \{z \in \mathbb{C} : |z| = 1\}$. By λ we shall denote the normalized one dimensional Lebesgue measure on ∂D ; by H^1 the linear subspace of the Banach space $L^1(\lambda, \partial D)$ which is the closure in the norm $\int_{\partial D} |f(z)| d\lambda$ of $A(D)$. In general if A is a uniform algebra on X and ρ is a positive Borel measure on X, then $H^1_A(\rho)$ denotes the linear subspace of $L^1(\rho, X)$ which is the closure in the norm $\int_X |f(x)| d\rho$ of functions of A regarded as elements of $L^1(\rho, X)$.

The natural injection $i_{A,\rho} : A \to H^1_A(\rho)$ is defined to be the map which corresponds to each $f \in A$ the same function regarded as an element of $H^1_A(\rho)$. Clearly $i_{A,\rho}$ is an absolutely summing operator because $\|i_{A,\rho}(f)\| = \int_X |f(x)| d\rho \le \int_X |f(x)| d\rho$ for every $f \in A$.

Our first result (Proposition 1) combined with the Criterion gives in fact the proof of Theorem 1 for the disc algebra because the natural injection $i_{A(D),\lambda}$ is clearly a non compact linear operator. Since H^1 is a dual Banach space, lemma 1 yields that every operator to H^1 which locally factors through L^1-spaces actually factors through some L^1-space.

<u>Proposition 1</u>. Every operator from the disc algebra to the Hardy space H^1 which factors through an L^1-space is compact.

<u>Proof.</u> It is enough to show

1° Every operator from an L^1-space into H^1 takes weakly compact sets into compact sets

2° Every operator from $A(D)$ into an L^1-space is weakly compact.

For 1° observe that H^1 is separable and, by the F. and M. Riesz Theorem, it is a dual Banach space (the predual is the quotient of

$C(\partial D)$ by the subspace $\{f \in A : \int_{\partial D} f(z)d\lambda = 0\}$). Now use the Dunford-Pettis Theorem (cf. [2] Chapt. VI).

For 2° note that because the adjoint of a weakly compact operator is weakly compact it suffices to show that every operator from a $C(K)$-space (in particular from a dual of an L^1-space) into $[A(D)]^*$ is weakly compact. To this end, by a result of [7], it is enough to show that no subspace of $[A(D)]^*$ is isomorphic to the space c_0. We have (cf. [8])

$$[A(D)]^* = \left(L^1(\lambda,\partial D)/H_0^1\right) \times V_{sing}$$

where V_{sing} is the space of finite Borel measures on ∂D which are singular with respect to λ and $H_0^1 = \{f \in H^1 : \int_{\partial D} f(z)d\lambda = 0\}$. Since V_{sing} is an L^1-space, every operator from c_0 into V_{sing} is compact (cf. [2] Chapt. VI). Therefore if E were a subspace of $[A(D)]^*$ isomorphic to c_0 then the restriction to E of the natural projection of $[A(D)]^*$ onto $L^1(\lambda,\partial D)/H_0^1$ would be a Fredholm operator. Hence $L^1(\lambda,\partial D)/H_0^1$ would contain a subspace, say E_1, isomorphic to c_0. Thus, by a result of Sobczyk (cf. e.g. [7]), E_1 would be complemented in $L^1(\lambda,\partial D)/H_0^1$ and therefore in $[A(D)]^*$. This would contradict the fact that no complemented subspace of a dual Banach space is isomorphic to c_0 (cf. [1]).

<u>Remark.</u> An inspection of the proof shows that if A is a uniform algebra on X with a norm separable annihilator in $[C(X)]^*$ and if ρ is a finite positive Borel measure on X such that every measure orthogonal to A is absolutely continuous with respect to ρ, then every operator from A into $H_A^1(\rho)$ which factors through an L^1-space is compact.

In the sequel we shall need the following concept

Definition. A triple $(\rho,(f_n),F)$ is called a Lebesgue transporter for the uniform algebra A on X if

(i) ρ is a normalized positive Borel measure on X

(ii) $f_n \in A$ and $\|f_n\| \leq 1$ for $n = 1, 2, \ldots$

(iii) $F(x) = \lim_n f_n(x)$ and $|F(x)| = 1$ for ρ-almost all $x \in X$

(iv) $F\rho \in A^{\perp}$ i.e. $\int_X f(x)F(x)d\rho = 0$ for all $f \in A$.

The next result is the main technical tool in the proof of Theorem 1.

Proposition 2. If $(\rho,(f_n),F)$ is a Lebesgue transporter for the uniform algebra A, then the natural injection $i_{A,\rho} : A \to H^1_A(\rho)$ does not factor locally through L^1-spaces.

Proof. Let $U_n : A(D) \to A$ be defined by $U_n(f) = f \circ f_n$ for $f \in A(D)$ $(n = 1, 2, \ldots)$.
Next define an operator $V : H^1_A(\rho) \to H^1$ as follows. Given $g \in H^1_A(\rho)$ let x^*_g be the unique functional on $C(\partial D)$ such that

$$x^*_g(f) = \int_X (f \circ F(x))g(x)d\rho \quad \text{for } f \in C(\partial D).$$

Let $e_m(z) = z^m$ for $z \in \partial D$ and $m = 0, \pm 1, \pm 2, \ldots$.
Then, by (iii) and (iv),

$$x^*_g(e_m) = \int_X F^m(x)g(x)d\rho = 0 \quad \text{for } m = 1, 2, \ldots$$

because if we choose a sequence (g_n) in A so that

$$\lim_n \|g_n - g\|_{H^1_A(\rho)} = 0$$

then

$$\int_X F^m(x)g(x)d\rho = \lim_n \int_X [f_n(x)]^{m-1} g_n(x)F(x)d\rho = 0.$$

Thus, by the F. and M. Riesz Theorem x_g^* corresponds via the Riesz Representation Theorem to a measure $h \cdot \lambda$ for some $h \in H^1$. We put

$$V(g) = h .$$

Now assume to the contrary that the operator $i_{A,\rho}$ factors locally through L^1-spaces. Let us consider the sequence of operators (K_n) defined by

$$K_n = Vi_{A,\rho}U_n : A(D) \to H_A^1(\rho) \quad (n= 1, 2, \ldots)$$

We have

$$\|K_n\| \leq \|V\| \, \|i_{A,\rho}\| \, \|U_n\| \leq 1 \quad \text{for} \quad n = 1, 2, \ldots .$$

By (iii) for every $f \in A(D)$ we have

$$\lim_{n,r\to\infty} \|i_{A,\rho}U_n(f) - i_{A,\rho}U_r(f)\| = \lim_{n,r\to\infty} \int_X |f \circ f_n(x) - f \circ f_r(x)| d\rho = 0$$

Therefore there exists a strong limit,

$$\lim_n K_n(f) = K(f) \quad \text{for} \quad f \in A(D).$$

Hence our assumption that $i_{A,\rho}$ factors locally through L^1-spaces implies, by Lemma 2, that $K : A(D) \to H^1$ has the same property. Thus K factors through an L^1-space because of Lemma 1 and the fact that H^1 is a dual Banach space.
Hence, by Proposition 1, K is a compact operator. The desired contradiction now follows from the identity

$$(*) \qquad K(e_m) = e_m \quad \text{for} \quad m = 1, 2, \ldots$$

because

$$\|e_m - e_r\|_{H^1} \geq 1 \quad \text{for} \quad m \neq r.$$

To check $(*)$ put $K_n(e_m) = h_{n,m}$. Then for $s = 0, \pm 1, \pm 2, \ldots$ we have

$$\int_{\partial D} h_{n,m}(z) z^s d\lambda = \int_X F^s(x) [f_n(x)]^m d\rho .$$

Hence

$$\binom{*}{*} \quad \lim_n \int_{\partial D} h_{n,m}(z) z^s d\lambda = \int_X F^{s+m}(x) d\rho = \begin{cases} 1 & \text{for} \quad s = -m \\ 0 & \text{for} \quad s \neq -m \end{cases}$$

(because, by (iii) and (iv), $\int_X F^s(x)d\rho = 0$ for $s > 0$ and by the positivity of the measure ρ and, by (iii),

$$\int_X F^s(x)d\rho = \overline{\int F^{-s}(x)d\rho} = 0 \quad \text{for} \quad s < 0).$$

Clearly $\binom{*}{*}$ implies (*).

Our last Proposition uses a standard uniform algebra technique and goes back to Bishop.

<u>Proposition 3</u>. Let A be a uniform algebra on a compact Hausdorff space X. Let us consider the following three properties

(1) there exists in the maximal ideal space $M(A)$ a Gleason part which contains at least to different points,

(2) there exists an ideal $J \subset A$ and a multiplicative linear functional φ on A such that $0 < a < 1$ where $a = \|\varphi|J\| = \sup_{f \in J} |\varphi(f)|$

(3) there exists a Lebesgue transporter for A

Then $(1) \Rightarrow (2) \Rightarrow (3)$.

<u>Proof</u> $(1) \rightarrow (2)$. If φ and Ψ are in the same Gleason part, then φ with $J = \ker \Psi$ satisfy (2) (cf. [4], Chapt. VI, Theorem 2.1).

$(2) \rightarrow (3)$. Let μ be a Borel measure on X which is a Hahn-Banach extension of $\varphi|J$ onto $C(X)$, i.e. $\int_X f(x)d\mu = \varphi(f)$ for $f \in J$ and $\|\mu\| = a$. Let $F = \frac{d|\mu|}{d\mu}$ be the Radon-Nikodym derivative of the total variation $|\mu|$ of μ with respect to μ.

Let $\rho_1 = |1-aF|^2|\mu|$ and $\rho = \rho_1/\|\rho_1\|$. Finally let (f_n) be a sequence in A such that

$$a = \lim_n |\varphi(f_n)| = \lim_n |\int_X f_n(x)d\mu|.$$

Then $(\rho,(f_n),F)$ is a Lebesgue transporter for A. The proof of this fact is implicite contained in the proof of Theorem 7.1 in Chapt. VI of [4].

Proof of Theorem 1. Combine the Criterion with Propositions 2 and 3 remembering that the natural injection $i_{A,\rho} : A \to H^1_A(\rho)$ is an absolutely summing operator.

Added in proof:
After this paper was submitted for publication, the author learned that a result similar to that of Kisliakov [6] had been obtained by Etcheberry [10].

REFERENCES.

[1] C. Bessaga and A. Pełczyński, On bases and unconditional convergence of series in Banach spaces, Studia Math. 17 (1958), 151 - 164.

[2] N. Dunford and J.T. Schwartz, Linear Operators I, Interscience, New York 1958.

[3] T. Figiel, W.B. Johnson and L. Tzafriri, J. Approximation Theory, 13 (1975), 395 - 412.

[4] T.W. Gamelin, Uniform algebras, Prentice Hall, Englewood Cliffs, N.J. 1969.

[5] Y. Gordon and D.R. Lewis, Absolutely summing operators and local unconditional structures, Acta Math. 133 (1974), 27 - 48.

[6] V.L. Kisliakov, Mat. Zametki, to appear (in Russian).

[7] A. Pełczyński, Projections in certain Banach spaces, Studia Math. 19 (1960).

[8] A. Pełczyński, On simultaneous extension of continuous functions, Studia Math., 24 (1964), 285 - 304.

[9] A. Pełczyński, Sur certaines propriétés nouvelles des espaces de Banach de fonctions holomorphes A et H^∞, Compt. Rend. Acad. Sci. Paris, t. 279, série A (1974), 9 - 12.

[10] A. Etcheberry, Some uncomplemented uniform algebras, Proc. Amer. Math. Soc., 43 (1973), 323 - 325.

ALGEBRAS BETWEEN L^{∞} AND H^{∞}

Donald Sarason

This report concerns the following theorem:

Let B be a closed subalgebra of L^{∞} (of the unit circle) which contains H^{∞} (the algebra of boundary functions for bounded holomorphic functions in the unit disk). Let B_I be the closed subalgebra of L^{∞} generated by H^{∞} and the complex conjugates of those inner functions that are invertible in B. Then $B = B_I$.

The possibility that this result might be true was originally advanced by R. G. Douglas in 1968. The question arose in connection with the study of Toeplitz operators [1].

The proof of the theorem breaks into two parts, both of which are substantial and interesting. The parts are due, respectively, to S.-Y. Chang and D. Marshall. As the results of Chang and Marshall already exist in preprint form [2], [3], this report will not attempt to present all the details of their proofs. Rather, I shall give a general discussion of the Chang-Marshall theorem and try to get across the main ideas in the proof.

The theorem can be regarded as an L^{∞}-replacement for the Wermer maximality theorem [4]. Let C be the algebra of continuous functions on the unit circle and A the usual disk algebra ($A = C \cap H^{\infty}$). Wermer's theorem states that C and A are the only closed subalgebras of C that contain A. Between L^{∞} and H^{∞}, on the other hand, there is room for many closed subalgebras. The Chang-Marshall theorem says that all such algebras are determined, in a certain sense, by H^{∞}. Wermer's theorem can be stated, in roundabout fashion, as follows: If B is a closed subalgebra of C which contains A properly, then the inner function z is invertible in B. The Chang-Marshall theorem says that if B is a closed algebra between L^{∞} and H^{∞}, then B makes so many inner functions invertible that their complex conjugates, together with H^{∞}, generate B. The preceding rephrasing of Wermer's theorem, while somewhat artificial, does dovetail nicely with a proof of the theorem due to K. Hoffman and I. M. Singer [5, p. 93]. The proof has two steps: (1) It is shown that if B is a closed algebra between C and A which does not make the function z invertible, then Lebesgue measure is multiplicative on B. (2) From the latter conclusion one easily infers that $B = A$. Roughly, step (2) corresponds to Chang's half of the Chang-Marshall theorem, and step (1) corresponds to Marshall's half. The Hoffman-Singer method also enables one to show that any closed subalgebra of L^{∞} that contains H^{∞} properly makes the function z invertible and so contains $H^{\infty} + C$ [5, p. 193]. That implies, incidentially, that L^{∞} and H^{∞} are the only weak-star closed subalgebras of L^{∞} that contain H^{∞}, another L^{∞}-replacement for Wermer's theorem.

A discussion of the status of the Douglas problem as of June, 1972, can be found in [6]. The main results at that time consisted of some nonobvious examples of algebras satisfying the Douglas condition; see [7], [8], [9].

Further discussion requires some more notation. The unit disk will be denoted by D and the unit circle by ∂D. The Gelfand space (space of multiplicative linear functionals) of a Banach algebra B will be denoted by $M(B)$. We identify the unit disk with an open subset of $M(H^{\infty})$ by letting each point of D correspond to the functional on H^{∞} of evaluation at that point. We identify $M(L^{\infty})$ with a closed subset of $M(H^{\infty})$ by letting each functional in $M(L^{\infty})$ correspond to its restriction to H^{∞}. With that identification, $M(L^{\infty})$ becomes the Shilov boundary of H^{∞}. Each functional φ in $M(H^{\infty})$ is represented by a unique regular Borel probability measure m_{φ} on $M(L^{\infty})$. Thus, for f in H^{∞}, the Gelfand transform of f, which we also denote by f, is given by

$$f(\varphi) = \int f \, dm_{\varphi} \qquad (\varphi \in M(H^{\infty})).$$

We use the preceding equality to define $f(\varphi)$ for an arbitrary f in L^{∞} and φ in $M(H^{\infty})$. This associates with each f in L^{∞} a continuous function on $M(H^{\infty})$. In D, the latter function is just the Poisson integral of f.

For B a closed algebra between L^{∞} and H^{∞}, we identify $M(B)$ with a closed subset of $M(H^{\infty})$ by letting each functional in $M(B)$ correspond to its restriction to H^{∞}. By the uniqueness of representing measures mentioned above, distinct functionals in $M(B)$ have distinct restrictions to H^{∞}, so this identification is one-to-one. An inner function b is invertible in B if and only if $|b| = 1$ on $M(B)$. Conversely, if B is the closed subalgebra of L^{∞} generated by H^{∞} and the complex conjugates of a family of inner functions (we refer to such an algebra as a Douglas algebra), then $M(B)$ consists of the set of points in $M(H^{\infty})$ where each inner function in the family has unit modulus. These observations together with the Chang-Marshall theorem imply that each closed algebra between L^{∞} and H^{∞} is uniquely determined by its Gelfand space: if B and B_1 are such alegbras and $M(B) = M(B_1)$, then $B = B_1$.

Douglas realized that the latter conclusion would be a consequence of the theorem he had proposed. He suggested that, as a possible way of gaining insight into his problem, one should try to prove the above statement for the special case $B = H^{\infty} + C$. The author succeeded in doing that in the summer of 1972 [10]. Subsequently, the same result for other special algebras B was obtained by T. Weight [11], S. Axler [12], and Chang [13]. These efforts culminated, in January of 1975, in

CHANG'S THEOREM. _If_ B _and_ B_1 _are closed algebras between_ L^{∞} _and_ H^{∞} _such that_ B _is a Douglas algebra and_ $M(B) = M(B_1)$, _then_ $B = B_1$.

With the proof of this theorem, the complete solution of the Douglas problem suddenly appeared tantalizingly near. Marshall provided the final step in March of 1975.

MARSHALL'S THEOREM. If B is a closed algebra between L^∞ and H^∞, then $M(B) = M(B_I)$.

The proof of Chang's theorem is an interesting application of some of the techniques used by C. Fefferman and E. M. Stein [14] in establishing their characterizations of the space BMO. For simplicity, we outline the proof of Chang's theorem for the special case where B is $H^\infty[\bar{b}]$, the closed algebra generated by H^∞ and the complex conjugate of the single inner function b. This special case contains all of the essential difficulties. We suppose that B_1 is a closed algebra between L^∞ and H^∞ such that $M(B) = M(B_1)$. Since b is obviously invertible in B_1 we have $B \subset B_1$, so it remains to establish the reverse inclusion.

In all that follows, we let K denote an absolute constant, possibly different on each occurance.

Choose any function w in B_1. Our aim is to show that w is in $H^\infty[\bar{b}]$. Adding a constant to w if necessary, we can assume without loss of generality that w is invertible in B_1. If we multiply w by the outer function whose modulus is $|w|^{-1}$, we obtain a unimodular invertible function in B_1, and the latter function will belong to B if and only if w does. Thus, we may assume, without loss of generality, that $|w| = 1$ almost everywhere on ∂D.

The algebra $H^{\infty}[\bar{b}]$ is spanned by the subspaces $\bar{b}^n H^{\infty}$, $n = 1, 2, \ldots$. Thus, we want to show that $\operatorname{dist}(w, \bar{b}^n H^{\infty}) \to 0$. Since b is unimodular, we have $\operatorname{dist}(w, \bar{b}^n H^{\infty}) = \operatorname{dist}(wb^n, H^{\infty})$. By a well-known duality principle, the distance of an L^{∞} function from H^{∞} equals the norm of the functional induced by the function on H^1_0. Hence, we want to estimate $\frac{1}{2\pi}\int_{-\pi}^{\pi} wb^n g\, dt$, where g belongs to H^1_0. For technical reasons (which will not emerge completely in the following sketch), we restrict our attention to functions g that are holomorphic in a neighborhood of $\bar{D}$. To estimate the above integral we replace it by an integral over the unit disk, using the following lemma.

LEMMA. _If_ u _and_ v _belong to_ L^2 _of the unit circle and one of_ u _and_ v _has mean value_ 0, _then_

$$\frac{1}{2\pi}\int_{-\pi}^{\pi} uv\, dt$$

$$= \frac{1}{\pi}\iint_D (\operatorname{grad} u)\cdot(\operatorname{grad} v) \log\frac{1}{|z|}\, dxdy.$$

In the preceding statement, the functions u and v are assumed to be extended harmonically into D by means of Poisson's formula.

Applying the lemma to the case at hand, with $u = w$ and $v = b^n g$, we obtain

$$\left| \frac{1}{2\pi} \int_{-\pi}^{\pi} wb^n g \, dt \right|$$

$$\leqq \frac{1}{\pi} \iint_D |\text{grad } w| |\text{grad } b^n g| \log \frac{1}{|z|} \, dxdy.$$

Since $b^n g$ is holomorphic in D, the absolute value of its gradient is equal to $2^{1/2}$ times the absolute value of its derivative. Thus

$$(*) \quad \left| \frac{1}{2\pi} \int_{-\pi}^{\pi} wb^n g \, dt \right|$$

$$\leqq \frac{2^{1/2}}{\pi} \iint_D |b|^n |\text{grad } w| |g'| \log \frac{1}{|z|} \, dxdy$$

$$+ \frac{2^{1/2} n}{\pi} \iint_D |b|^{n-1} |g| |\text{grad } w| |b'| \log \frac{1}{|z|} \, dxdy.$$

Now the function w, being unimodular and invertible in B_1, has unit modulus on $M(B_1)$ and hence on $M(B)$. But $M(B)$ is the set of points in $M(H^\infty)$ where $|b| = 1$. Thus, $|w| = 1$ everywhere in $M(H^\infty)$ where $|b| = 1$. So, $|w|$ must be close to 1 everywhere in $M(H^\infty)$ (and, in particular, everywhere in D) where $|b|$ is close to 1. The preceding property of w is the one that enables us to obtain favorable estimates of the integrals on the right side of $(*)$. We fix an ε in $(0, 1)$ and let G denote the subset of D where $|w| \geqq 1-\varepsilon$ and (for technical reasons) where $|z| \geqq 1/2$. By the preceding observation we have $|b| \leqq 1-\delta$ on $D - G$ for some δ in $(0, 1)$.

We accomplish the desired estimates by using the technique of Feffermen and Stein from [14]. The technique depends (in a way we shall not make entirely explicit) upon a theorem of Carleson (see, for example, [15, p. 157]). For I a closed subarc of ∂D, let

$$R(I) = \{ re^{i\theta} : e^{i\theta} \in I, 0 < 1-r \leqq |I| \},$$

where $|I|$ denotes the normalized Lebesgue measure of I. For μ a finite positive measure on ∂D, let

$$\gamma(\mu) = \sup_I \mu(R(I))/|I|.$$

In case $\gamma(\mu) < \infty$ one calls μ a Carleson measure. The theorem of Carleson says, basically, that for all h in H^2 one has the inequality

$$\int_D |h|^2 \, d\mu \leqq K\gamma(\mu)\|h\|_2^2.$$

Following Fefferman and Stein, we associate with each function v in L^1 a measure μ_v on D, defined by

$$d\mu_v = (1-|z|)|\text{grad } v|^2 dxdy,$$

it being understood that v is extended into D via Poisson's formula. Fefferman and Stein in [14] establish inequalities (i) $\|v\|_* \leqq K\gamma(\mu_v)^{1/2}$ and (ii) $\gamma(\mu_v)^{1/2} \leqq K\|v\|_*$, where $\|v\|_*$ is the BMO norm of v. In particular, for v in L^∞ one has

(iii) $\gamma(\mu_v)^{1/2} \leq K\|v\|_\infty$.

Chang's basic lemma states that the measure associated with w is small on the set G:

LEMMA. $\gamma(\chi_G \mu_w) \leq K\varepsilon$.

The proof resembles the Fefferman-Stein proof of (ii). Armed with this lemma, one attacks the integrals on the right side of (*) using the Fefferman-Stein method for establishing (i). The measure μ_w makes its appearance after an application of Schwarz's inequality. On G the above lemma together with (iii) yields an estimate of the desired form. On $D - G$ one uses the inequality $|b| \leq 1-\delta$ and (iii). The outcome, after considerable juggling, is an inequality

$$\left| \frac{1}{2\pi} \int_{-\pi}^{\pi} b^n wg \, dt \right|$$

$$\leq K[\varepsilon^{1/2} + (1-\delta)^{n/2} + n(1-\delta)^{n-1}]\|g\|_1 .$$

From this one obtains $\operatorname{dist}(w, B) \leq K\varepsilon^{1/2}$, and Chang's theorem (for the special case under consideration) follows.

As indicated above, the general case of Chang's theorem requires only minor modifications.

Turning to Marshall's theorem, we again consider a special case to which the general case is easily reduced. Let $B = H^\infty[w]$, the closed algebra generated by H^∞ and a single L^∞ function w. As in the proof of Chang's theorem, we loose no generality by assuming that w is unimodular and invertible in B. Let B_1 be any closed algebra between H^∞ and B such that $M(B_1)$ properly contains $M(B)$. Marshall's method produces an interpolating Blaschke product which is invertible in B but not in B_1. Accordingly, one obtains a stronger theorem than was originally stated: _Every closed algebra between L^∞ and H^∞ is generated by H^∞ and the complex conjugates of interpolating Blaschke products_.

S. Ziskind [16] proved in 1974 that L^∞ itself is generated by H^∞ and the complex conjugates of interpolating Blaschke products. Marshall's proof is a (highly nontrivial) modification of Ziskind's. Both proofs are based on the central construction in L. Carleson's proof of the corona theorem [17].

As is easily verified, $M(B)$ consists of the set of points in $M(H^\infty)$ where $|w| = 1$. Since, by assumption, $M(B_1)$ properly contains $M(B)$, there is a point φ in $M(B_1)$ where $|w(\varphi)| < \alpha < 1$. The Marshall construction yields an interpolating Blaschke product b with the following properties:

(i) $|w| \leqq \beta < 1$ everywhere in D where $b = 0$;

(ii) $|b| \leqq 1/10$ everywhere in D where $|w| < \alpha$.

(The number β appearing in (i) depends only on α.) Since b is an interpolating Blaschke product, its zero set in $M(H^\infty)$ is the closure of its zero set in D. Hence, it follows from (i) that $|w| \leqq \beta$ wherever b vanishes in $M(H^\infty)$. In particular, then, b does not vanish on $M(B)$, so b is invertible in B. On the other hand, by the corona theorem, the set of points in $M(H^{\infty})$ where $|w| < \alpha$ is contained in the closure of its intersection with D. Hence, by (ii), $|b| \leqq 1/10$ everywhere in $M(H^{\infty})$ where $|w| < \alpha$. In particular, $|b(\varphi)| \neq 1$, so b is not invertible in B_1.

To obtain b, Marshall constructs a system Γ of contours in $\bar{D}$ surrounding the subset of D where $|w| < \alpha$ such that $|w| \leqq \beta$ on Γ and such that arc length measure on Γ is a Carleson measure. The construction of Γ is complicated; it is based on the construction Carleson invented to solve the corona problem [17]. On $\Gamma \cap D$ one distributes a sequence $(z_n)_1^{\infty}$ of points (which are to be the zeros of the desired function b) such that

(1) $|(z_m - z_n)/(1-\bar{z}_n z_m)| > \varepsilon > 0$ for $m \neq n$;

(2) for each z in $\Gamma \cap D$ there is an n such that $|(z-z_n)/(1-\bar{z}_n z)| < 1/10$.

From (1) and the fact that arc length measure on Γ is a Carleson measure, one infers that the measure $\Sigma_1^{\infty} (1-|z_n|^2)\delta_{z_n}$ is also a Carleson measure. That, in the presence of (1), is enough to guarantee that (z_n) is an interpolating sequence. From (2) one concludes that $|b| \leqq 1/10$ on $\Gamma \cap D$. The construction of Γ is carried out in such a way that $\Gamma \cap D$ has harmonic measure 0 relative to each component of the interior of Γ. One can thus conclude that $|b| \leqq 1/10$ on the interior of Γ and, in particular, on the set where $|w| < \alpha$, thereby verifying condition (ii). Condition (i) holds because the zeros of b are on Γ, which is contained in the set where $|w| \leq \beta$.

There remain a number of interesting problems concerning closed algebras between L^∞ and H^∞.

STRUCTURE. Let B be a closed algebra between L^∞ and H^∞. Let C_B be the C^*-algebra generated by the inner functions that are invertible in B. It has been verified in a number of cases that B is equal to the linear span of H^∞ and C_B [8], [10], [13]. Is it true in general that $B = H^\infty + C_B$?

MEAN OSCILLATION. Let Q_B be the largest C^*-algebra contained in B. In several cases the functions in Q_B can be characterized in terms of mean oscillation [10], [12], [13]. Is there a general theorem along these lines? The methods of Chang should be effective here.

APPROXIMATION BY QUOTIENTS. Can every unimodular function in C_B be uniformly approximated by quotients of inner functions that are invertible in B? Again, a positive answer is known in several cases [8], [13]. It is tempting to try to obtain an abstract approximation theorem which would answer this question.

SETS OF ANTISYMMETRY. Suppose f is a function in L^∞ such that $f|\text{supp}\, m_\varphi$ is in $B|\text{supp}\, m_\varphi$ for every φ in $M(B)$. Then m_φ is multiplicative on the algebra $B[f]$ for each φ in $M(B)$, from which one concludes that $M(B) = M(B[f])$. Hence, by the Chang-Marshall theorem, $B[f] = B$, in other words, f must belong to B. This result should be compared to the conclusion one gets from Bishop's theorem on sets of antisymmetry [18, p. 60]: if $f|S$ is in $B|S$ for each maximal set of antisymmetry S of B in $M(L^\infty)$, then f is in B. Because the support sets of representing measures are sets of antisymmetry, the former conclusion is the stronger of the two. There should, one suspects, be a link between the former conclusion and Bishop's theorem, even though no such link is suggested by the proof of the Chang-Marshall theorem. A natural question is whether every maximal set of antisymmetry is made up, in a "nice way," of support sets. Apparently, no theorem along these lines is possible for a general function algebra, but one might be true in the present situation.

REFERENCES

1. R. G. Douglas, On the spectrum of Toeplitz and Wiener-Hopf operators, Abstract Spaces and Approximation (Proc. Conf. Oberwohlfach, 1968), Birkhäuser, Basel, 1969, pp. 53-66.

2. S.-Y. A. Chang, A characterization of Douglas subalgebras, preprint.

3. D. E. Marshall, Subalgebras of L^∞ containing H^∞, preprint.

4. J. Wermer, On algebras of continuous functions, Proc. Amer. Math. Soc. 4 (1953), 866-869.

5. K. Hoffman, Banach Spaces of Analytic Functions, Prentice-Hall, Englewood Cliffs, N. J., 1962.

6. D. Sarason, Algebras of functions on the unit circle, Bull. Amer. Math. Soc. 79 (1973), 286-299.

7. R. G. Douglas and W. Rudin, Approximation by inner functions, Pacific J. Math. 31 (1969), 313-320.

8. A. M. Davie, T. W. Gamelin and J. Garnett, Distance estimates and pointwise bounded density, Trans. Amer. Math. Soc. 175 (1973), 37-68.

9. D. Sarason, Approximation of piecewise continuous functions by quotients of bounded analytic functions, Canadian J. Math. 24 (1972), 642-657.

10. D. Sarason, Functions of vanishing mean oscillation, Trans. Amer. Math. Soc., to appear.

11. T. Weight, Some subalgebras of L^∞ determined by their maximal ideal spaces, Bull. Amer. Math. Soc. 81 (1975), 192-194.

12. S. Axler, Some properties of $H^\infty + L_E^\infty$, unpublished.

13. S.-Y. Chang, On the structure and characterization of some Douglas subalgebras, Amer. J. Math., to appear.

14. C. Fefferman and E. M. Stein, H^p spaces of several variables, Acta Math. 129 (1972), 137-193.

15. P. L. Duren, The Theory of H^p Spaces, Academic Press, New York, 1970.

16. S. Ziskind, Interpolating sequences and the Shilov boundary, to appear.

17. L. Carleson, Interpolations by bounded analytic functions and the corona problem, Ann. Math. 76 (1972), 547-559.

18. T. W. Gamelin, Uniform Algebras, Prentice-Hall, Englewood Cliffs, N. J., 1969.

THE MODULUS OF CONTINUITY OF AN ANALYTIC FUNCTION

H.S. Shapiro

1.1 Introduction.

The theme of my talk today is "elementary mathematics from an advanced standpoint" and will consist of some observations concerning the following theorem (D shall always denote the open unit disc):

Theorem 1. *Suppose f is continuous in D^- (closure of D) and analytic in D, and define for $a > 0$*

$$\omega(a) = \omega(a;f) = \sup_{\substack{z_1, z_2 \in D^- \\ |z_1 - z_2| \le a}} |f(z_2) - f(z_1)|$$

and

$$\tilde{\omega}(a) = \tilde{\omega}(a;f) = \sup_{\substack{z_1, z_2 \in \partial D \\ |z_1 - z_2| \le a}} |f(z_2) - f(z_1)|$$

Then, there is an absolute constant C such that

$$(1.1) \quad \omega(a) \le C\,\tilde{\omega}(a).$$

This theorem is due to Tamrazov [6], and another proof may be found in Rubel, Shields and Taylor [3]. It is known that the constant C in (1.1) cannot taken to be 1. (The reverse inequality to (1.1), with $C = 1$, is of course trivial).

It is interesting to contrast Theorem 1 with the following result of Hardy and Littlewood [2]:

Theorem 2. *Suppose f is continuous in D^- and harmonic in D. With ω, $\tilde{\omega}$ as above,*

$$(1.2) \quad \omega(a) \le C\,\tilde{\omega}(a) \log \frac{1}{a}, \qquad a \le \frac{1}{2}$$

where C is an absolute constant.

The log factor in (1.2) cannot be dropped. Thus, an essentially better theorem holds for analytic than for merely harmonic functions. I would like to sketch proofs of these theorems from the standpoint of harmonic analysis, based on ideas I have expounded earlier [4,5], and which have been improved upon in an essential way by Jan Boman [1]. I believe this method of proof affords insight into the theorems ; also it yields various generalizations.

Let us turn, then, to Theorem 1. By the change of variables $f(e^{it}) = F(t)$ we can pass to a function F analytic in the half-plane $\operatorname{Im} t > 0$, which is more convenient for our method. Supposing this done, and retaining the notation $f(z)$ for our function, we assume then: f <u>is continuous in</u> P^-, <u>the closure of the upper half-plane</u> P, <u>and analytic in</u> P. Before we can attack the problem by our method proper, we need some preliminary simplifications.

Let f_y denote the "slice function" $x \to f(x+iy)$ on $\mathbb{R}$, where $y \geq 0$. Then, since f_y is the convolution of f_0 with a probability measure (Poisson kernel) it follows at once that its modulus of continuity is majorized by that of f_0 which, by definition, is $\tilde{\omega}$.

Next, let $z_1 = x_1 + iy_1$ and $z_2 = x_2 + iy_2$ be any two points of P^-, where we may suppose $y_1 \leq y_2$. Let $a = |z_1 - z_2|$; then

$$|f(z_2)-f(z_1)| \leq |f(z_2)-f(z_3)| + |f(z_3)-f(z_1)|$$

where $z_3 = x_2 + iy_1$. The second term on the right does not exceed the modulus of continuity of f_{y_1}, evaluated at $|z_3-z_1| = |x_2-x_1|$, and so, by the above remark is $\leq \tilde{\omega}(|x_2-x_1|) \leq \tilde{\omega}(a)$. We shall, therefore, have proved the theorem if we show that $|f(z_2)-f(z_3)|$ does not exceed some absolute constant times the modulus of continuity of f_{y_1}, evaluated at $|z_2-z_3| = |x_2-x_3| \leq a$. And this clearly follows from:

<u>Proposition</u>. <u>Let</u> f <u>be continuous in</u> P^- <u>and analytic in</u> P. <u>There is an absolute constant</u> C <u>such that</u>, <u>for</u> $a > 0$

$$(1.3) \qquad |f(x+ia)-f(x)| \leq C\,\tilde{\omega}(a)$$

<u>where</u>

$$(1.4) \qquad \tilde{\omega}(a) = \tilde{\omega}(a;f) = \sup_{\substack{x_1,x_2 \in \mathbb{R} \\ |x_1-x_2| \leq a}} |f(x_1)-f(x_2)|.$$

Here we interrupt our discussion for some preliminaries.

1.2. Dilations of measures and generalized m.o.c.

Let $M(\mathbb{R})$ denote the Banach algebra of bounded complex measures on $\mathbb{R}$, considered in the standard way as the conjugate space of $C_0(\mathbb{R})$. The dilation $x \to ax$, for each positive a, induces an isometry on $C_0(\mathbb{R})$ whose adjoint is an isometry on $M(\mathbb{R})$; the image of $\mu \in M(\mathbb{R})$ under this isometry we call the a-dilation of μ, denoted by $\mu_{(a)}$. It is easy to see, denoting Fourier transform by $^\wedge$, that $\hat{\mu}_{(a)}(x) = \hat{\mu}(ax)$.

To each $\mu \in M(\mathbb{R})$ corresponds a generalized modulus of continuity (m.o.c) of any bounded uniformly continuous function f on $\mathbb{R}$, defined by

$$\omega_\mu(a;f) = \sup_{0<b\leq a} \|f*\mu_{(b)}\| .$$

Here $\|\cdot\|$ denotes sup norm on $\mathbb{R}$, and $*$ denotes convolution. In these terms observe that $\tilde{\omega}(a;f)$ as defined by (1.4) is the same thing as $\omega_\sigma(a;f_0)$ where σ is the difference of two point masses,

$$(1.5)\quad \sigma = \delta_0 - \delta_1 , \quad \hat{\sigma}(x) = 1 - e^{-ix}.$$

Using the Poisson representation of f we see also that $f(x) - f(x+ia) = f_0 * \tau_{(a)}$ where

$$(1.6)\quad \tau = \delta_0 - \frac{dx}{\pi(1+x^2)} , \quad \hat{\tau}(x) = 1 - e^{-|x|}.$$

With these notations, (1.3) can be written (writing henceforth simply f in place of f_0)

$$(1.7)\quad \omega_\tau(a;f) \lesssim C\, \omega_\sigma(a;f).$$

It is in this form that we shall prove the theorem.

1.3. Conclusion of the proof.

It is a simple exercise [4] that (1.7) will hold between any two measures σ, τ if σ divides τ in $M(\mathbb{R})$. Is this the case here? That is, does the ratio $\hat{\tau}(x)/\hat{\sigma}(x)$ belong to FM (Fourier transforms of bounded measures)? Obviously not. First of all $\hat{\sigma}(x)$ has zeros at all the points $2\pi n$, $|n| \geq 1$ and $\hat{\tau}(x)$ does not vanish there. Furthermore, although both $\hat{\tau}(x)$ and $\hat{\sigma}(x)$ vanish at $x = 0$ their ratio has a jump discontinuity there, and so certainly is not the restriction of any function in FM to a neighbourhood of 0. We must, to conclude the proof, show how each of these difficulties can be overcome.

The first step is based on an idea of Boman [1], indeed we could quote a general theorem of his, but in this simple case we prefer to proceed quite directly. We shall namely replace σ by a "smeared" version σ^* whose Fourier transform is nowhere small away from the origin. Let us define

$$(1.8) \quad \sigma^* = \int_0^1 \sigma_{(t)}\, dt$$

where the integral is understood e.g. in Bochner's sense.

Observe that $\sigma^*_{(a)} = \int_0^1 \sigma_{(at)}\, dt$, and so

$\|f*\sigma^*_{(a)}\| \leq \int_0^1 \|f*\sigma_{(at)}\| dt \leq \omega_\sigma(a;f)$. Hence

$$(1.9) \quad \omega_{\sigma^*}(a;f) \leq \omega_\sigma(a;f)$$

and so it is enough to prove (1.7) with ω_{σ^*} in place of ω_σ. (An inequality in the opposite direction to (1.9) is also valid, although we do not require it here ; thus the generalized moduli of continuity determined by the "naive" measure σ and the "sophisticated" one σ^* mutually dominate one another).

From (1.8) we see that

$$\widehat{\sigma^*}(x) = \int_0^1 \hat{\sigma}_{(t)}(x)dt = \int_0^1 (1-e^{-it\,x})dt$$

$$= \frac{e^{-ix} - 1 + ix}{ix}.$$

Observe that $\widehat{\sigma^*}$ is bounded away from zero outside any neighbourhood of the origin.

Now things would be nice if $\hat{\tau}(x)/\widehat{\sigma^*}(x)$ coincided with an element of FM near $x = 0$; this is not so, however, since $\widehat{\sigma^*}(x)$ behaves like a constant times x. If our f were merely continuous we could, indeed, push the analysis no further at this point, and this singularity at $x = 0$ is, in a sense, the reason why Theorem 2 contains a coarser estimate than Theorem 1. However, our f is the boundary value of an analytic function on P. In terms of harmonic analysis this means that its spectrum lies on the positive half line. Hence $\int f \, d\mu = 0$ for every $\mu \in M(\mathbb{R})$ whose Fourier transform is supported on the positive half line ("analytic measure"). This implies that we may replace τ by another measure τ^* and still have $\omega_{\tau^*}(a;f) = \omega_{\tau}(a;f)$, providing only $\widehat{\tau^*}(x) = \hat{\tau}(x)$ for $x < 0$. In particular we may define

$$\widehat{\tau^*}(x) = \begin{cases} 1 - e^{x} & , \quad -\infty < x \leq 1 \\ 1 - e^{-x} & , \quad x \geq 2 \\ \text{linear} & , \quad 1 \leq x \leq 2 \end{cases}$$

Clearly $\tau^* \in M(\mathbb{R})$. Also $\widehat{\tau^*}(x) \big/ \widehat{\sigma^*}(x)$ is C^1 and bounded on the whole real line. To be sure that it is the Fourier transform of something in $M(\mathbb{R})$ we have, finally, only to check the neighbourhood of infinity, i.e. to make sure the "Wiener-Pitt phenomenon" does not occur. This is very easily done by direct examinations ; alternatively, we may quote a general theorem of Varopoulos [7] which (specialized to $\mathbb{R}^1$) implies that the closed subalgebra M_0 of M generated by all discrete and all absolutely continuous measures has a maximal ideal space in which $\hat{\mathbb{R}}$ is dense. Since σ^* is the sum of a discrete and an absolutely continuous measure, $\sigma^* \in M_0$. Since $|\widehat{\sigma^*}(x)| \geq c > 0$ for $|x| \geq 1$, we conclude that $1 \big/ \widehat{\sigma^*}(x)$ coincides, on $|x| \geq 1$, with the Fourier transform of a bounded measure on $\mathbb{R}$. Thus, finally, $\widehat{\tau^*} \big/ \widehat{\sigma^*}$ belongs to FM and therefore $\omega_{\tau^*}(f;a) \leq C_1 \, \omega_{\sigma^*}(f;a)$ for some absolute constant C_1. Putting it all together, we have Theorem 1.

1.4. Remarks.

As already pointed out, the "smearing" step (i.e. passing to σ^*) has been carried out in a very general situation by Boman [1] ; also Theorem 2 follows from Boman's results even more directly, since now there is no need (or possibility!) to pass from τ to τ^*.

2. Some generalizations.

Of course, the above proof of Theorem 1 is longer than the extant proofs. Its advantage lies in being part of a general scheme, which yields various generalizations with little extra effort. Perhaps the most striking of these arises upon replacing τ by

$$\tau_0 = \delta_0 - k(x)dx$$

where

$$k(x) = \frac{1}{2\pi}\left(\frac{\sin t/2}{t/2}\right)^2, \qquad \hat{k}(x) = (1-|x|)^+.$$

Then

$$\hat{\tau}_0(x) = \begin{cases} |x|, & |x| \le 1 \\ 1, & |x| \ge 1 \end{cases}$$

whereupon the analysis is essentially identical with the preceding. In case f has period 2π, it is easily verified that $f * (\tau_0)_{(1/n)}$ equals the classical Fejér sum σ_{n-1} and we obtain

<u>Theorem 3.</u> <u>If</u> f <u>is continuous in</u> D^- <u>and analytic in</u> D, <u>its Fejer sum of order</u> n <u>satisfies</u>

$$(2.1) \qquad \|f-\sigma_n\| \le C\,\tilde{\omega}(1/n), \qquad n \ge 1$$

<u>where</u> C <u>is an absolute constant, and</u> $\tilde{\omega}$ <u>is as in Theorem 1</u>.

It would be rash to claim that such an apparently elementary result was not known earlier ; in any case I have not encountered it elsewhere. Of course, if f is merely continuous on the circle (without the analyticity) it is well known that (2.1) holds with an essential factor of $\log(n+1)$ on the right. Thus, a "Jackson" estimate is actually provided <u>by the Fejér sums, for "analytic"</u> f, i.e. those which have one-sided Fourier series.

Another source of generalization is to use a different norm in place of the sup norm. This yields (placed in the context of the circle):

Theorem 4. For $1 \leq p < \infty$, we have

$$(2.2) \quad \left(\frac{1}{2\pi}\int_0^{2\pi} |f(re^{i\theta}) - f(e^{i\theta})|^p \, d\theta\right)^{1/p} \leq C\tilde{\omega}_p(1-r)$$

for all $f \in H^p(D)$, where C is an absolute constant. Here $\tilde{\omega}_p(a)$ denotes the function

$$\sup_{0<b\leq a} \left(\frac{1}{2\pi}\int_0^{2\pi} |f(e^{i(\theta+b)}) - f(e^{i\theta})|^p \, d\theta\right)^{1/p}$$

(so-called L^p modulus of continuity).

It may be observed that for $1 < p < \infty$, (2.2) is valid even for merely harmonic f in D. This is because $\hat{\tau}(x) / \hat{\sigma}(x)$ in the above proof, while not belonging to FM near $x = 0$ does coincide with a Fourier p-multiplier there. (This is essentially the M. Riesz Theorem). The case $p = 1$ requires holomorphy, and is presumably new.

Finally we remark that essentially the same method of proof extends Theorem 2 to the polydisc. We indicate very summarily the necessary changes in the proof (for notational simplicity, treating the case of two variables). We assume first that the problem has been reduced to the form of "Proposition" above.

Next, if $a = (a_1,a_2)$ is any pair of positive numbers, the a-dilation of $\mu \in M(\mathbb{R}^2)$ is defined as above, so that $\hat{\mu}_{(a)}(x) = \hat{\mu}_{(a)}(x_1,x_2) = \hat{\mu}(a_1x_1,a_2x_2)$. If σ denotes the Dirac measure at $(0,0)$ minus the Dirac measure at $(1,1)$, then for f bounded and continuous on $\mathbb{R}^2$ we have $f * \sigma_{(a)} = f(x) - f(x-a)$ and so

$$\omega_\sigma(t) = \sup_{\|a\|\leq t} \|f * \sigma_{(a)}\|$$

is the ordinary m.o.c. of f. In like manner the quantity we want to estimate is $\omega_\tau(t)$, where τ is the Dirac measure at $(0,0)$ minus $k\,dx$,

$$k(x) = \pi^{-2}(1+x_1{}^2)^{-1}(1+x_2{}^2)^{-1}.$$

The Fourier transforms of the measures we must now work with are, then

$$\hat{\sigma}(x) = 1 - e^{-i(x_1+x_2)}$$

$$\hat{\tau}(x) = 1 - e^{-|x_1+x_2|}$$

We introduce the "smeared" measure

$$\sigma^* = \int_0^1\int_0^1 \sigma_{(a)}\, da_1 da_2$$

so that

$$\widehat{\sigma^*}(x_1 x_2) = 1 - \int_0^1\int_0^1 e^{-i(a_1x_1+a_2x_2)}\, da_1 da_2$$

is small only near $x_1 = x_2 = 0$, and moreover behaves like a constant times $x_1 + x_2$ near the origin. Finally, since the holomorphy of f is equivalent to f having its spectrum in the first quadrant of $\hat{\mathbb{R}}^2$, $\int f\, d\mu = 0$ whenever $\mu \in M(\mathbb{R}^2)$ satisfies $\hat{\mu}(-x) = 0$ for all x in the first quadrant. Thus, $\omega_\tau(t,f) = \omega_{\tau^*}(t,f)$ if the Fourier transform of τ^* agrees with that of τ on the third quadrant. It is easy to see that there exists a function $\widehat{\tau^*}$ equal to $\hat{\tau}(x)$ on the third quadrant, and moreover equal to $1 - e^{(x_1+x_2)}$ on a full neighbourhood of $(0,0)$. In fact, simply define $\widehat{\tau^*}(x) - \hat{\tau}(x)$ to be the function $e^{-|x_1|-|x_2|} - e^{x_1+x_2}$, multiplied by anything in FM of compact support which equals 1 on a neighbourhood of the origin. Then σ^* divides τ^* in $M(\mathbb{R}^2)$, and the proof is concluded like that of Theorem 1.

REFERENCES

[1] Jan Boman, On comparison theorems for generalized moduli of continuity, in Linear Operators and Approximation Theory (Proceedings of a conference at Oberwolfach 1974), P.L. Butzer and B. Sz-Nagy, editors, pages 105 - 111.

[2] G.H. Hardy and J.E. Littlewood, "Some properties of fractional integrals, II" Math. Zeit. 34 (1931) 403 - 439.

[3] L.A. Rubel, A.L. Shields and B.A. Taylor, "Mergelyan sets and the modulus of continuity of analytic functions", preprint.

[4] H.S. Shapiro, Smoothing and approximation of functions, Van Nostrand, New York, 1969.

[5] H.S. Shapiro, Topics in approximation theory, Springer lecture notes in math., vol. 187, 1971.

[6] P.M. Tamrazov, Contour and solid structure properties of holomorphic functions of a complex variable, Russ. Math. Surveys 28 (1973) 141 - 173 ; see also other papers of Tamrazov referred to therein.

[7] N.Th. Varopoulos, Studies in harmonic analysis, Proc. Cambridge Philo. Soc. 60 (1964) 465 - 516.

Multi-dimensional analytic structure in the spectrum of a uniform algebra

by Nessim Sibony

We are interested in giving conditions to yield several dimensional analytic structure in the maximal ideal space $\mathfrak{M}$ of a uniform algebra A. Most of the known results are one dimensional, see [11] and [12].

We shall consider first a specific maximum principle satisfied by holomorphic and plurisubharmonic functions in $\mathbb{C}^n$. Generalizing this maximum principle to a uniform algebra we consider the Shilov boundaries introduced by R. Basener in [1]. After considering some basic examples, we give generalisations of classical results in several complex variables such as the maximum principle for varieties and the Hartogs' principle. Then we prove a several variables version of Wermer's maximality theorem, theorem 8 below.

Using a theorem of R. Basener [1], which generalize a result of Bishop [2], we give conditions which yield Stein spaces in the maximal ideal space of a uniform algebra. We then give applications to polynomially convex hulls of compact sets in $\mathbb{C}^n$, theorems 17 and 18. The basic tools are some inequalities from geometric measure theory from [4] and [9].

We also give a necessary and sufficient condition for the existence of analytic sets of dimension k in $\mathfrak{M}$. The condition is in terms of Hausdorff measures associated with the Gleason metric on $\mathfrak{M}$, theorem 20.

This paper was developped independently of Basener's work [1] cited above. Since some of the results (essentially theorem 11) have appeared in [1], we only state them for later use and refer to Basener's paper for their proof.

It is a pleasure for me to thank Monique Hakim for her helpful comments during the preparation of this work.

NOTATIONS.

We recall here some definitions and notations. If A is a uniform algebra on a compact Hausdorff space X, We denote by $\mathfrak{M}$ or $\mathfrak{M}_A$ its maximal ideal space. For every $V \subset \mathfrak{M}$ and $f \in A$, f_V is the restriction of f to V and $\|f\|_V = \sup\{|f(x)|, x \in V\}$. If F is a closed set in $\mathfrak{M}$ then A_F is the function algebra generated in $\mathscr{C}(F)$ by f_F, for f in A ; $h(F) = \{x \in \mathfrak{M} / |f(x)| \leq \|f\|_F\}$, we shall say that F is A-convex if $h(F) = F$.

When Ω is an open domain in $\mathbb{C}^n$, $A(\bar{\Omega})$ will denote the algebra of functions continuous in $\bar{\Omega}$ and holomorphic in Ω, $\mathscr{H}(\Omega)$ will denote the Frechet algebra of functions holomorphic in Ω. If K is a compact set in $\mathbb{C}^n$ we write $H(K)$ for the closure in $\mathscr{C}(K)$ of the functions holomorphic in a neighborhood of K and we write $P(K)$ for the closure of polynomials. D will denote the unit disc in $\mathbb{C}$, $B(z_0, r)$ the euclidian ball in $\mathbb{C}^n$ of center x_0 and radius r and $S(z_0, r)$ the corresponding sphere. We shall use Stout's book [11] as a reference in uniform algebra theory.

1. A MAXIMUM PRINCIPLE.

Let Ω be a bounded domain in $\mathbb{C}^n$, $n \geq 2$, let $p \in \Omega$ and let h be a non-constant holomorphic map from D into Ω with $h(0) = p$. We then say that $\Delta_p = h(D)$ is an analytic disc through p. We shall say that a function ψ is harmonic (respectively subharmonic) in Δ_p, if $\psi \circ h$ is harmonic (respectively subharmonic) in D. Write $\mathcal{S}(\Omega)$ for the cone of continuous functions ψ in Ω such that for every $p \in \Omega$ there is an analytic disc, Δ_p, through p, such that ψ is harmonic on Δ_p.

THEOREM 1. Let Ω be a bounded domain in $\mathbb{C}^n$, and let φ and ψ be continuous functions in $\bar{\Omega}$ with φ plurisubharmonic in Ω and $\psi \in \mathcal{S}(\Omega)$. Suppose we have

$$\varphi(z) \leq \psi(z) \quad \text{for every} \quad z \in \partial\Omega.$$

Then,

$$\varphi(z) \leq \psi(z) \quad \text{for every} \quad z \in \bar{\Omega}.$$

Proof. Let M be the maximum of $\varphi - \psi$ in $\bar{\Omega}$; suppose that $M > 0$. Denote by K the compact set

$$K = \left\{ z \in \bar{\Omega} / (\varphi - \psi)(z) = M \right\}.$$

Clearly K is contained in Ω. Let $p \in K$ be a peak point for the algebra $H(K)$. Let Δ_p be the analytic disc through p associated with the function ψ. The function $\varphi - \psi$ is subharmonic in Δ_p and has its maximum in p. Hence $\varphi - \psi$ is constant in Δ_p and therefore $\Delta_p \subset K$. Since p is a peak point for $H(K)$, Δ_p cannot be contained in K and so $M \leq 0$. The theorem is proved.

EXAMPLES. 1) Let $(g_1, \ldots, g_{n-1})$ be an $(n-1)$-tuple of functions in $A(\bar{\Omega})$ and θ a continuous function in $\mathbb{C}^{n-1}$. Then the function

$$\psi = \theta(g_1, \dots, g_{n-1})$$

is in $\mathcal{S}(\Omega)$. Let $p \in \Omega$ and define

$$V_p = \{z \in \Omega, \quad g_i(z) = g_i(p) \quad 1 \le i \le n-1\}$$

V_p is an analytic variety and so there is an analytic disc through p in V_p. Observe that the restriction of ψ to V_p is constant. In particular if $f, g_1, \dots, g_{n-1}$

$$\operatorname{Re} f \le \sum_{i=1}^{n-1} |g_i| \quad \text{on} \quad \partial\Omega$$

then

$$\operatorname{Re} f \le \sum_{i=1}^{n-1} |g_i| \quad \text{on} \quad \bar{\Omega}.$$

This will be our starting point in introducing Shilov boundaries of different orders.

2) Let $(f_1, \dots, f_n)$ be an n-tuple of functions in $A(\bar{\Omega})$ and $(\alpha_1, \dots, \alpha_u)$ an u-tuple of positive numbers. Suppose that the function

$$u = \sum_{i=1}^{n} |f_i|^{\alpha_i}$$

does not vanish in Ω. Then, the function $\log u$ is in $\mathcal{S}(\Omega)$. We shall prove this. Let $p \in \Omega$ and suppose $f_1(p) \ne 0$. Then, in a neighborhood U of p there are holomorphic functions $g_1, \dots, g_n$, such that $f_1^{\alpha_1} = g_i^{\alpha_i}$ for $2 \le i \le n$. Hence

$$\log u = \log |f_1|^{\alpha_1} + \log(1 + \left|\frac{f_2}{g_2}\right|^{\alpha_1} + \dots + \left|\frac{f_n}{g_n}\right|^{\alpha_n}).$$

Define

$$V_p = \{z \in U \mid \; f_i/g_i(z) = f_i/g_i(p) \qquad 2 \le i \le n\}.$$

Then V_p is an analytic set in U and therefore contains an analytic disc through p, moreover $\log u$ is harmonic on V_p.

COROLLARY 2. Let φ be a continuous function in an open set $\Omega \subset \mathbb{C}^n$. The following condition is necessary and sufficient for φ to be plurishbharmonic.

If B is a closed ball in Ω and $p_0, p_1, \dots, p_{n-1}$ are n analytic polynomials such that

$$\varphi \le \operatorname{Re} p_0 + \sum_{i=1}^{n-1} |p_i| \quad \text{on} \quad \partial B$$

it follows that

$$\varphi \le \operatorname{Re} p_0 + \sum_{i=1}^{n-1} |p_i| \quad \text{in} \quad B$$

Proof. The necessity of the condition is clear from theorem 1 and example 1. To prove that the condition is sufficient repeat the argument of i) $\Rightarrow$ ii) in theorem 3 below.

2. SHILOV BOUNDARY OF ORDER k.

As before, A will denote a uniform algebra on a compact space X with maximal ideal space $\mathfrak{M}$, and k will be a positive number. If $G = (g_1, \ldots, g_k) \in A^k$, let

$$V(G) = \{m \in \mathfrak{M} / G(m) = 0\}.$$

THEOREM 3. Let F_k be a closed subset of $\mathfrak{M}$. The following properties are equivalent :

i) For every $G = (g_1, \ldots, g_k) \in A^k$ and $f \in A$

$\operatorname{Re} f \le \sum_{i=1}^{k} |g_i|$ on F_k implies $\operatorname{Re} f \le \sum_{i=1}^{k} |g_i|$ on $\mathfrak{M}$.

ii) For every $G \in A^k$ we have

a) $V(G) \cap F_k = \emptyset$ implies $V(G) = \emptyset$

b) If $h \in A$ and $|h| \le 1$ on $V(G) \cap F_k$, then $|h| \le 1$ on $V(G)$.

Proof. i) $\Rightarrow$ ii). a) Suppose that $V(G) \cap F_k = \emptyset$. Then, there exists $\varepsilon > 0$ such that $\sum_{i=1}^{k} |g_i| \ge \varepsilon$ on $V(G) \cap F_k$; using i) it follows that $\sum_{i=1}^{k} |g_i| \ge \varepsilon$ on $\mathfrak{M}$. Hence $V(G) = \emptyset$.

b) Let $h \in A$ and $|h| \le 1$ on $V(G) \cap F_k$. Fix $\varepsilon > 0$. There is a neighborhood V_ε of $V(G) \cap F_k$ such that on V_ε $|h| \le 1 + \varepsilon$. Therefore if C_ε is big enough

$$* \qquad |h| \le (1+\varepsilon) + C_\varepsilon \sum_{i=1}^{k} |g_i| \quad \text{on} \quad F_k.$$

From i) we deduce that $*$ holds in $\mathfrak{M}$ and so on $V(G)$ we have

$$|h| \le (1+\varepsilon).$$

Since ε was arbitrary we conclude $|h| \le 1$.

ii) $\Rightarrow$ i). Suppose that

$$\operatorname{Re} f \le \sum_{i=1}^{k} |g_i| \quad \text{on} \quad F_k.$$

If m_0 is a point in $\mathfrak{M}$, define

$$Z_{m_0} = \{m \in \mathfrak{M} / g_i(m) = g_i(m_0), \quad 1 \le i \le k\}.$$

Since $m_0 \in Z_{m_0}$, ii) a) implies that $Z_{m_0} \cap F_k \ne \emptyset$. But on $Z_{m_0} \cap F_k$

$$\operatorname{Re} f(m) \le \sum_{i=1}^{k} |g_i(m)| = \sum_{i=1}^{k} |g_i(m_0)|.$$

Hence, by ii) b), the same inequality holds in Z_{m_0}, and so

$$\operatorname{Re} f(m_0) \le \sum_{i=1}^{k} |g_i(m_0)|.$$

Since m_0 was arbitrary the implication is proved

If F_k satisfies i) or ii), we see that for every $G \in A^k$, F_k contains the Shilov boundary of $A_{V(G)}$. Therefore F_k contains the set Σ_k, sometimes denoted $\Sigma_k(A)$, defined by

$$(1) \qquad \Sigma_k = \text{closure}\left[\bigcup_{G \in A^k} \operatorname{Sh} A_{V(G)}\right].$$

On the other hand, it is clear that Σ_k satisfies ii). Thus we can make the following definition.

DEFINITION. The Shilov boundary of order k, of a uniform algebra A is the smallest closed subset F_k in $\mathfrak{M}$ satisfying the conditions of theorem 3.

It is clear that this set is Σ_k and that Σ_0 is just the usual Shilov boundary.

To check that a set F_k contains Σ_k we need only show that F_k satisfies property (1) for $G \in \mathcal{A}^k$ for some dense subset $\mathcal{A}$ of A.

EXAMPLES. α) Let K be a fat holomorphically convex compact set in $\mathbb{C}^n$, i. e. the maximal ideal space of $H(K)$ can be identified with K. If Σ_k is the Shilov boundary of order k associated with $H(K)$, we have

$$\Sigma_0 \subset \Sigma_1 \subset \ldots \subset \Sigma_{n-1} \subset \partial K \quad \text{and} \quad \Sigma_n = K.$$

In fact, if $k \leq n-1$ and

$$V(G) = \{z \in K / g_1(z) = \ldots = g_k(z) = 0, \quad g_i \in H(K)\}$$

then $V(G) \cap \overset{o}{K}$ is an analytic set and each branch of $V(G)$ goes to the boundary of K. Consequently ii) a) is satisfied with $F_k = \partial K$; ii) b) is the maximum principle for analytic varieties, see [6] p. 106.

β) Let D^n be the unit polydisc in $\mathbb{C}^n$ and $A = A(\bar{D}^n)$. Then

$$\Sigma_i = \{z = (z_1, \ldots, z_n) / n\text{-}i \text{ components of } z \text{ are of modulus } 1\}.$$

The proof is left to the reader.

γ) The following example is more delicate. Let Ω be a pseudoconvex domain in $\mathbb{C}^n$, with smooth boundary. That is there exists a $\mathcal{C}^\infty$ function ρ in a neighborhood U of $\bar{\Omega}$ such that

$$\Omega = \{z \in U / \rho(z) < 0\},$$

and $\operatorname{grad} \rho \neq 0$ on $\partial\Omega$. We recall that Ω is pseudoconvex if and only if

$$\langle \mathcal{L}(\rho)(z)w, w\rangle = \sum_{j,k=1}^{h} \frac{\partial^2 \rho(z)}{\partial z_j \partial \bar{z}_k} w_j \bar{w}_k \geq 0$$

when $z \in \partial\Omega$ and $\sum_{j=1}^{n} \frac{\partial \rho}{\partial z_j} w_j = 0$.

We shall regard the Levi form $\mathcal{L}(\rho)(z)$ as acting only on the complex tangent space to $\partial\Omega$ at z. Define

$$N_k = \{z \in \partial\Omega / \dim \ker \mathcal{L}(\rho)(z) \leq k\}.$$

We recall that a compact set is an S_δ set if it is an intersection of Stein domains, and then it is holomorphically convex [8].

PROPOSITION 4. Let Ω be a smooth pseudoconvex set in $\mathbb{C}^n$. Suppose that $\bar{\Omega}$ is an S_δ set. Then, the Shilov boundary of order k, $k \leq n-1$ associated with $H(\bar{\Omega})$ is the closure of N_k.

Proof. Suppose firstly that $k = 0$ and so N_0 is just the set of strictly pseudoconvex points, the theorem is simply that of Rossi [8] p. 489 characterising the Shilov boundary of $H(\bar{\Omega})$. In fact $\bar{N}_0 = \Sigma_0$ even when Ω is in a Stein manifold see [8].

We give a proof for $k = 1$. Since $\bar{\Omega}$ is an S_δ we can write $\bar{\Omega} = \bigcap_p \Omega_p$, each Ω_p being a domain of holomorphy. Let $g \in \mathcal{H}(\Omega_p)$ and $f \in H(\bar{\Omega})$. Suppose that $\operatorname{Re} f \leq |g|$ on $\bar{N}_1$. We shall prove that $\operatorname{Re} f \leq |g|$ on $\bar{\Omega}$. If g is constant, the assertion is obvious, using the case $k = 0$. Suppose then that g is non constant and observe that, by Hartogs' theorem $g(\Omega) \subset g(\partial\Omega)$. Let g_1 be the restriction of g to the manifold $\partial\Omega$. By Sard's theorem the set of critical values of g_1 is of measure zero in $\mathbb{C}$; but the set of critical values of g considered as a mapping from Ω_p to $\mathbb{C}$ is also of measure zero. Hence almost every point in $g(\Omega)$ is a regular value of g and g_1. Define

$$W = \left\{\alpha \in \mathbb{C} / \alpha \text{ regular value for } g \text{ and } g_1\right\}.$$

Fix $\alpha \in W$ and let $V_\alpha = g^{-1}(\alpha)$. Since α is a regular value for g, V_α is a manifold in Ω_p and it is clear that V_α is a Stein manifold : moreover since α is a regular value for g_1 $V_\alpha \cap \partial\Omega$ is a submanifold of $\partial\Omega$ and so $V_\alpha \cap \Omega$ is an open set of V_α with smooth boundary. But, if $z_0 \in \partial\Omega \cap V_\alpha$ is a strictly pseudoconvex point on V_α it is clear that $z_o \in N_1$. Consequently, since $\mathrm{Re}\, f \leq |\alpha|$ on N_1, using the result for $k = 0$ cited above, we conclude that $\mathrm{Re}\, f \leq |\alpha|$ on $V_\alpha \cap \Omega$. We have thus proved that

$$\mathrm{Re}\, f \leq |g| \quad \text{on} \quad g^{-1}(W).$$

Since $g(\Omega)\setminus W$ is of measure zero it is easy to see that $g^{-1}(W)$ is dense in Ω, observe that g is an open map. Therefore

$$\mathrm{Re}\, f \leq |g| \quad \text{on} \quad \bar{\Omega}.$$

Since $\bigcup_p \mathcal{H}(\Omega_p)$ is dense in $H(\bar{\Omega})$, it is clear that N_1 satisfies property i) of theorem 3.

The proof for general k follows the same lines using an induction argument.

We now give a generalization of the known fact that if $f \in H(B)$, where B is the unit ball in $\mathbb{C}^n$, $n \geq 2$, and if $|f| = 1$ on ∂B, then f is constant.

PROPOSITION 5. Let A be a uniform algebra with maximal ideal space , and let f, g be two functions in A. Suppose that $|f| = |g|$ on Σ_1. Then $|f| = |g|$ on $\mathfrak{M}$.

If A is antisymmetric and if $|f| = 1$ on Σ_1 then f is constant.

Proof. Let $\alpha \in \mathbb{C}$, $|\alpha| = 1$. It is clear that

$$\mathrm{Re}(\alpha f) \leq |g| \quad \text{on} \quad \Sigma_1.$$

Then we have

$$\mathrm{Re}\, \alpha f \leq |g| \quad \text{on} \quad \mathfrak{M}.$$

Since α was arbitrary it follows that $|f| \leq |g|$ on $\mathfrak{M}$. Reversing the rôle of f and g, we see that $|f| = |g|$ on $\mathfrak{M}$.

Recall that a uniform algebra is antisymmetric if and only if the only real-valued function in A are the constants. If $|f| = 1$ on Σ_1, we have seen that $|f| = 1$

on $\mathfrak{M}$, and hence f is invertible with $f^{-1} = \bar{f}$. Thus $2\,\mathrm{Re}\, f = f + \bar{f} = f + f^{-1}$ is in A and hence constant, $2\,\mathrm{Im}\, f = -i(f - \bar{f})$ is also constant. Therefore f is constant.

In this context there is a natural generalization of Rossi's maximum modulus principle.

PROPOSITION 6. [1]. Let K be a compact A-convex subset of $\mathfrak{M}$. Then

$$\Sigma_k(A|_K) \subset \partial K \cup (\Sigma_k \cap K).$$

For a proof see [1], p. 99.

We shall now prove an abstract maximum principle for A-varieties. Let $(A, \mathfrak{M})$ be a uniform algebra and let N be an open set of $\mathfrak{M}$. Recall that a function f is in $\mathcal{H}_A(N)$ if at every point y of N there is an open set W, $y \in W$, such that f is uniformly approximable on W by elements of A, see [11]. A closed subset V of $\mathfrak{M}$ is called an A-variety of codimension less than k if for every $x \in V$ there is an open set N, $x \in N$, such that

$$V \cap N = \{m \in N \;\; f_1(m) = \ldots = f_k(m) = 0, \;\; f_i \in \mathcal{H}_A(N) \;\; 1 \leq i \leq k\}.$$

PROPOSITION 7 (Hartogs' principle). Let V be an A-variety of codimension less than k. Then every component of V meets Σ_k and for every $f \in A$

$$\|f\|_V = \|f\|_{V \cap \Sigma_k}.$$

Proof. It follows from a theorem of Rickart (see [11] p. 361) that V is A-convex. Let $x_0 \in V \setminus \Sigma_k$ and let U be an A-convex neighborhood of x such that $U \cap \Sigma_k = \emptyset$, and such that

$$V \cap U = \{m \in U, \; f_1(m) = \ldots = f_k(m) = 0 \;\; f_i \in \mathcal{H}_A(U)\}.$$

Restricting U we can assume that f_i, $1 \le i \le k$, are in A_U. Proposition 6 implies that $\Sigma_k(A_U) \subset \partial U$. Hence the Shilov boundary of $A_{V\cap U}$ is in $V \cap \partial U$ and therefore

$$|f(x_o)| \le \|f\|_{V\cap\partial U}.$$

Consequently, the Shilov boundary of A_V is contained in $V \cap \Sigma_k$.

Using the Shilov idempotent theorem we see that each component of V meets Σ_k.

We now give a version of Wermer's maximality theorem.

THEOREM 8. Let K be a compact polynomially convex set in $\mathbb{C}^n$ with connected interior. Let B be a closed subalgebra of $\mathcal{C}(\partial K)$ containing the polynomials and whose Shilov boundary of order $n-1$ is identified with ∂K. Then either

$$B \subset A(K) \text{ and } \mathfrak{M}_B \text{ is identified with } K, \text{ or}$$

$$\mathfrak{M}_B \text{ is identified with } \partial K.$$

Proof. We need the following lemma which is implicit in the literature.

LEMMA 9. Let X be a polynomially convex set in $\mathbb{C}$, and let B be a uniform algebra on ∂X containing the function z. Define $\pi : \mathfrak{M}_B \to \mathbb{C}$ by

$$\pi(m) = m(z).$$

If $\pi(\mathfrak{M}_B) = X$ then $B = P(X)$; if $\pi(\mathfrak{M}_B) = \partial X$ then $\mathfrak{M}_B$ is isomorphic with ∂X.

Proof. Since X is polynomially convex we have that $\pi(\mathfrak{M}_B) \subset X$. The algebra generated by the polynomials is Dirichlet in $\mathcal{C}(\partial X)$, so if $\lambda \in \partial X$ then $\pi^{-1}(\lambda)$ reduces to the evaluation at λ. Therefore ∂X can be identified with

$\pi^{-1}(\partial X)$. Hence the lemma is proved if $\pi(\partial X) = \partial X$.

Suppose that $\pi(\mathfrak{M}) = X$. Let λ be a point in $\overset{\circ}{X}$ and let μ_1, μ_2 be measures on ∂X representing two homomorphisms $m_1, m_2 \in \pi^{-1}(\lambda)$. Since for every polynomial P we have $m_i(P) = P(\lambda)$, $i = 1, 2$, and since $P(\partial X)$ is Dirichlet it follows that $\mu_1 = \mu_2$ and so $m_1 = m_2$. Thus the map π is injective and since $\pi(\mathfrak{M}_B) = X$, $\mathfrak{M}_B$ can be identified with X. Using the local maximum principle as in [12] p. 62, we see that the functions in B are holomorphic in X and by Mergelyan's theorem $B = P(X)$.

LEMMA 10. [1]. <u>Let $k \geq 0$ and let $F \in A^k$. If W is a component of $\mathbb{C}^n \setminus F(\Sigma_{k-1})$, then either $F(\mathfrak{M}) \cap W = \emptyset$ or $F(\mathfrak{M}) \cap W = W$.</u> For a proof see [1] lemma 2.

Proof of theorem 8. Write $z_1, \ldots, z_n$ for the coordinate functions in $\mathbb{C}^n$. Define $\pi: \mathfrak{M}_B \to \mathbb{C}$ by

$$\pi(m) = (m(z_1), \ldots, m(z_n)).$$

Since $\overset{\circ}{K}$ is connected, lemma 10 implies that either $\pi(\mathfrak{M}_B) = K$ or $\pi(\mathfrak{M}_B) = \partial K$.

Suppose that $\pi(\mathfrak{M}_B) = K$ and let $\alpha = (\alpha_1, \ldots, \alpha_n) \in K$. Denote by $\mathfrak{M}_a^i$ the compact set

$$\mathfrak{M}_\alpha^i = \{m \in \mathfrak{M}_B \quad m(z_j) = \alpha_j \quad j \neq i\}.$$

Using the hypothesis $\Sigma_{n-1} \approx \partial K$, we see that the Shilov boundary of $B_{\mathfrak{M}_\alpha^i}$ can be identified with

$$S_\alpha^i = \{z \in \partial K \ , \ z_j = \alpha_j \ , \ j \neq i\}.$$

Let $\tilde{S}_\alpha^i$ denote the compact set

$$\tilde{S}_\alpha^i = \{z \in K, \quad z_j = \alpha_j \ , \ j \neq i\}.$$

The algebra B restricted to S_α^i contains z_i and its spectrum projects on $\tilde{S}_\alpha^i$, so by lemma 9, $\mathfrak{M}_{\alpha}^{i}$ is isomorphic with S_{α}^{i}, and π is injective. Moreover, $B \subset A(K)$, by the theorem on separate analyticity.

If $\pi(\mathfrak{M}_B) = \partial K$, We have seen that $\mathfrak{M}_{\alpha}^{i}$ is isomorphic with S_{α}^{i}, so $\mathfrak{M}_B$ is isomorphic with ∂K.

3. ANALYTIC STRUCTURE IN THE MAXIMAL IDEAL SPACE.

We recall here, for later use, a theorem proved by R. Basener [1]. As before A will denote a uniform algebra whose maximal ideal space is denoted by $\mathfrak{M}$.

THEOREM 11. [1]. Let $F \in A^n$ and let W be a component of $\mathbb{C}^n | F(\Sigma_{n-1})$. Assume that $F(\mathfrak{M}) \cap W \neq \emptyset$. Suppose $\exists W' \subseteq W$ such that $m_{2n}(W') > 0$ (m_{2n} is the Lebesgue measure in $\mathbb{C}^n$) and $\forall z \in W'$,

$$\# F^{-1}(z) = (\text{number of } m \in \mathfrak{M} \text{ with } F(m) = z)$$

is finite. For $\ell = 1, 2, \ldots,$ set

$$W_\ell = \{ z \in W / \# F^{-1}(z) = \ell \}.$$

Then there exists a positive integer k such that

(i) $W = \bigcup_{j=1}^{k} W_j$;

(ii) $\bigcup_{j=1}^{k-1} W_j$ is a proper analytic subvariety of W ;

(iii) $S = (F^{-1}(W), F, W)$ is a branched analytic cover, consequently $F^{-1}(W)$ is an analytic space and for every $f \in A$, f is holomorphic on $F^{-1}(W)$.

When $n = 1$ this theorem is due to Bishop, see [2] and [12]. For $n > 1$ the proof follows along the same lines, see [1].

We now give an addendum to theorem 11, which is of interest only when $n > 1$.

THEOREM 12. Suppose that the hypotheses of theorem 11 are satisfied and suppose that the envelope of holomorphy of W is a finitely sheeted Riemann domain. Then the analytic space $(S, F^{-1}(W), W)$ extends to a Stein space $\tilde{S}$ in $\mathfrak{M}$ and the functions of A are holomorphic in $\tilde{S}$.

Remark that in general $\tilde{S}$ is not open in $\mathfrak{M}$; consider for example the algebra A of continuous functions in the unit ball in $\mathbb{C}^2$ which are holomorphic in $\frac{1}{2} < |z| < 1$. If $F = (z_1, z_2)$, then

$$F(\Sigma_1) = \left\{ z \in C^2 / |z| \leq \tfrac{1}{2} \quad \text{or} \quad |z| = 1 \right\}$$

and clearly $\tilde{S}$ is isomorphic with the unit ball and $\tilde{S}$ is not an open set in $\mathfrak{M}$.

Proof of theorem 12. We denote by $S_\ell(A)$ the space of homogeneous polynomials of degree ℓ, on A. Recall that if P is a polynomial of degree ℓ there exists a unique ℓ-linear symmetric form on A, $\tilde{P}$, such that for every $x \in A$

$$P(x) = \tilde{P}(x, \ldots, x).$$

Since $\tilde{P}$ is unique we shall identify sometimes P with $\tilde{P}$. Remark that $S_\ell(A)$ has a natural Banach space topology, see [3].

We shall use the following lemmas.

LEMMA 13. Let φ be a map defined on a complex manifold Ω into $S_\ell(A)$. Suppose that for every $a \in A$ the map $z \to \varphi(z)a$ is analytic. Then φ is analytic.

Proof. If $\ell = 1$ then $S_\ell(A) = A'$; the lemma is proved by Gleason in [5] p. 513. For $\ell \geq 1$, consider $\tilde{\varphi}$ the map associated with φ ; $\tilde{\varphi}$ has its values in the Banach space of ℓ-linear mappings on A. Observe that if $P \in S_\ell(A)$ and $\tilde{P}$ is the associated multilinear map then if

$$\|P(x)\| \leq M \quad \text{when} \quad \|x\| \leq 1$$

hence

$$\|\tilde{P}(x_1,\ldots,x_\ell)\| \leq \frac{2^\ell}{\ell!} M \qquad \text{for} \qquad \|x_i\| \leq \frac{1}{\ell}, \quad 1 \leq i \leq n. \quad [3].$$

Using this and the Banach Steinhaus theorem we prove that $\tilde{\varphi}$ is locally bounded. The proof can be finished as for the case $\ell = 1$ [5] using power series expansion and Cauchy formula.

LEMMA 14. Let f be a holomorphic map defined on an open set $W \subset \mathbb{C}^n$ into a Banach space E and let $\tilde{W}$ be the envelope of holomorphy of W. Then f extends to a holomorphic maps from $\tilde{W}$ into E.

Proof. Let $H(W,E)$ denote the Frechet space of holomorphic functions with values in E. This space is isomorphic with the ε-tensor product $H(W)\hat{\otimes}_\varepsilon E$ and $H(W)$ is isomorphic with $H(\tilde{W})$, then $H(W,E)$ is isomorphic with $H(\tilde{W},E)$.

We prove now the theorem. For simplicity, we assume that $\tilde{W}$, the envelope of holomorphy of W, is an open set in $\mathbb{C}^n$. We use the notations of theorem 11. When $z \in W_k$, let $p_1(z), \ldots, p_k(z)$ be the k distinct points in $F^{-1}(z)$, and let $\sigma_1, \ldots, \sigma_k$ be the symmetric functions in $p_1, \ldots, p_k$; more precisely

$$\sigma_h(z) = \sum_{1 \leq i_1 < \ldots < i_h \leq k} p_{i_1}(z) \otimes p_{i_2}(z) \otimes \ldots \otimes p_{i_h}(z).$$

It is clear that $\sigma_h(z) \in S_h(A)$. For every $f \in A$, $z \to \sigma_h(z) f$ is holomorphic (use theorem 11). Then, by lemma 13, σ_h is a holomorphic function in W_k with values in $S_h(A)$. The function σ_h extends to a holomorphic function in W : in fact if (z_j) tends to $z_0 \in W \setminus W_k$ the sequence $\sigma_n(z_j)$ has cluster points, but, since for every $g \in A$, $z \to \sigma_n(z) g$ extends to a holomorphic function in W (note that $\sigma(z)\, g$ is bounded for $g \in A$ and use Riemann extension theorem), then there is only one cluster point, and lemma 13 shows that the extension of $z \to \sigma_h(z)$ to W is holomorphic.

Denote by $\Delta(z)$ the polynomial

$$\Delta(z) = \begin{cases} \bigotimes\limits_{\substack{i,j \\ i\neq j}} (p_i(z) - p_j(z)), & z \in W_k \\ 0, & z \in W \ W_k. \end{cases}$$

Δ is holomorphic in W with values in $S_\ell(A)$ with $\ell = n(n-1)$.

Using lemma 14, we see that the functions $\Delta, \sigma_1, \ldots \sigma_k$ extend to functions $\tilde{\Delta}, \tilde{\sigma}_1, \ldots, \tilde{\sigma}_k$ holomorphic in $\tilde{W}$ with values in the appropriate Banach space.

Denote by τ the map from $\mathfrak{M}^k$ to $\bigoplus\limits_{i=1}^{k} S_i(A)$ defined by

$$\tau(m_1, \ldots, m_k) = \sum_{i=1}^{k} \Big(\sum_{1 \le j_1 < j_2 \ldots < j_i \le k} m_{j_1} \otimes m_{j_2} \otimes \ldots \otimes m_{j_i} \Big).$$

<u>Assertion</u> 1. For every $z \in \tilde{W}$, $\tilde{\sigma}(z) = (\tilde{\sigma}_1(z), \ldots, \tilde{\sigma}_k(z))$ is a point in $\tau(\mathfrak{M}^k)$.

Proof of assertion 1. Observe that $\tau(\mathfrak{M}^k)$ is defined by algebraic equations in $\bigoplus\limits_{i=1}^{k} S_i(A)$. For example, if $k = 2$ (p,q) is in $\tau(\mathfrak{M}^2)$ if and only if for all x,y in A

$$\begin{cases} q(x,y) = q(x,1)\, q(1,y) \\ q(xy,1) = q(x,1)\, q(y,1) \\ q(1,xy) = q(1,x)\, q(1,y) \\ p(x) = q(x,1) + q(1,y). \end{cases}$$

Since $\tilde{\sigma}(W_k)$ is in $\tau(\mathfrak{M}^k)$ and this set is an analytic set, then, by analytic continuation, $\tilde{\sigma}(\tilde{W})$ is also contained in $\tau(\mathfrak{M}^k)$.

Denote by π_i the i-projection of $\mathfrak{M}^k$ on $\mathfrak{M}$. Observing that τ is injective define $S' = \tau^{-1}(\tilde{\sigma}(\tilde{W}))$. Let $S_i = \pi_i(S')$ and $\tilde{S} = \bigcup\limits_{i=1}^{k} S_i$.

We shall prove that $\tilde{S}$ has the structure of a Stein space. We first prove that

$(\tilde{S}, F, \tilde{W})$ is a branched analytic cover (see [6] p. 101) for the definition.

Clearly $F^{-1}(W) \subset \tilde{S} \subset \mathfrak{M}$. Let $\tilde{p}_i(z) = \pi_i \circ \tau^{-1} \circ \tilde{\sigma}(z)$.

Assertion 2. $F(\tilde{S}) = \tilde{W}$ and $(F \circ \tilde{p}_i)(z) = z$ for every $z \in \tilde{W}$.

Proof of assertion 2. Recall that $F = (f_1, \ldots, f_n)$. When $z \in W$, $\tilde{\sigma}_i(z) f_j$, $1 \le j \le k$, are the symmetric functions associated with

$$* \quad (p_1(z) f_j, \ldots, p_k(z) f_j) = (z_j, \ldots, z_j).$$

Since $z \to \tilde{\sigma}_i(z) f_j$ is holomorphic in $\tilde{W}$ relation $*$ remains true for $z \in \tilde{W}$ because $f_j \circ \tilde{p}_i(z)$ are by construction the roots of the equation

$$w^k - (\tilde{\sigma}_1(z) f_j) w^{k-1} + \ldots + (-1)^k (\tilde{\sigma}_k(z) f_j) = 0.$$

Therefore, for every $z \in \tilde{W}$, $1 \le i \le k$, $F \circ \tilde{p}_i(z) = z$. This proves the assertion.

Denote by $\tilde{V}$ the analytic subset of $\tilde{W}$ defined by

$$\tilde{V} = \left\{ z \in \tilde{W} / \tilde{\Delta}(z) = 0 \ \text{ i. e. } \ \tilde{\Delta}(z).g = 0 \ \forall g \in A \right\}.$$

Assertion 3. For every $z \in \tilde{W} \setminus \tilde{V}$ the points $\tilde{p}_1(z), \ldots, \tilde{p}_k(z)$ are distinct, and F is a covering map from $\tilde{S} \cap F^{-1}(\tilde{W} \setminus \tilde{V})$ to $\tilde{W} \setminus \tilde{V}$.

Proof of assertion 3. Observe that $F^{-1}(z) \cap \tilde{S}$ cannot contain more than k points. If $z \notin \tilde{V}$, there exists $g \in A$ with $\tilde{\Delta}(z)\, g \neq 0$ and so the equation

$$w^k - \tilde{\sigma}_1(z).g \ w^{k-1} + \ldots + (-1)^k \tilde{\sigma}_k(z).g = 0$$

has k distinct solutions. Since $\tilde{p}_1(z)\, g, \ldots, \tilde{p}_k(z)\, g$ are solutions it follows that $\tilde{p}_1(z), \ldots, \tilde{p}_k(z)$ are distinct.

Denote by Δ_k the generalized diagonal of $\mathfrak{M}^k$

$$\Delta_k = \left\{ (m_1, \ldots, m_k) \in \mathfrak{M}^k \ \exists\, i, j,\ i \neq j \ \text{ and } \ m_i = m_j \right\}.$$

If $z_0 \notin \tilde{V}$ then $\tau^{-1}(\tilde{\sigma}(z_0)) \notin \Delta_k$, and there exists a neighborhood U of $\tau^{-1}(\tilde{\sigma}(z_0))$ in $\mathfrak{M}^k$ such that $U \cap \Delta_k = \emptyset$. The map τ is continuous with respect to the weak topologies, since $\mathfrak{M}^k$ is compact and τ injective, the map $z \to \tau^{-1} \circ \tilde{\sigma}(z)$ is continuous. Thus there is an open set ω, $w \in z_0 \subset \tilde{W}$, such that $\tau^{-1} \circ \sigma(\omega) \subset U$. Therefore $\tilde{p}_i(z)$, $1 \leq i \leq k$, are distinct for $z \in \omega$ and the maps $z \to \tilde{p}_i(z) = \pi_i \circ \tau^{-1} \circ \tilde{\sigma}(z)$ are continuous in ω. Since $F \circ \tilde{p}_i(z) = z$ each $\tilde{p}_i$ is injective in ω and so it is a local homeomorphism. One see also that for every $g \in A$, $z \to \tilde{p}_i(z).g$ is holomorphic in ω.

Assertion 4. $(\tilde{S}, F, \tilde{W})$ is a branched analytic cover. (See [6] p. 101 for the definition).

Proof of assertion 4. The map F restricted to $\tilde{S}$ is proper : since, if K is a compact set in $\tilde{W}$, then $\tilde{S} \cap F^{-1}(K) = \bigcup_{i=1}^{k} \pi_i(\tau^{-1}(\tilde{\sigma}(K)))$. We now deduce that $\tilde{S}$ is locally compact when we give to $\tilde{S}$ the topology induced by the weak topology of $\mathfrak{M}$. If ω is an open set $\omega \subset\subset \tilde{W}$, then $\tilde{S}$ is locally the intersection of an open and a compact set : $F^{-1}(\omega) \cap (\tilde{S} \cap F^{-1}(\bar{\omega}))$, hence $\tilde{S}$ is locally compact.

It remains to show that $\tilde{S} \cap F^{-1}(\tilde{W} \setminus \tilde{V})$ is dense in $\tilde{S}$. Let m be a point in $F^{-1}(z_0) \cap \tilde{S}$ with $z_0 \in \tilde{V}$, and let $g \in A$ be such that $g(m_0) = 0$ and g separates the points of $F^{-1}(z_0) \cap \tilde{S}$. Choose a sequence (z_j) in $\tilde{W} \setminus \tilde{V}$ converging to z_0. Since the root system of a monic polynomial is a continuous of the coefficients, there exists $m_j \in \tilde{S}$ such that $F(m_j) = z_j$ and $\lim_{j \to \infty} m_j(g) = 0$. Since F is proper a subsequence of (m_j) converge to a point $m' \in \tilde{S}$, and $F(m') = z_0$, $m'(g) = 0$ so $m' = m_0$.

This completes the proof that $(\tilde{S}, F, \tilde{W})$ is an analytic cover and hence by a theorem of Remmert and Grauert [7] an analytic space. Since the $\tilde{\sigma}(z).f$, $f \in A$, are analytic in $\tilde{W}$ and satisfy a polynomial equation, then the function of A are analytic in $\tilde{S}$. Hence the space $H(\tilde{S})$ of holomorphic functions on $\tilde{S}$ separates the points of $\tilde{S}$. As F is proper and $\tilde{W}$ is Stein then $\tilde{S}$ is holomorphically convex ; therefore $\tilde{S}$ is a Stein space. This completes the proof.

When $\tilde{W}$ is a Riemann domain the proof is a little more technical but uses the same ideas.

4. APPLICATIONS.

We give now some application to the existence of analytic structure in the polynomially convex hull of a compact set in $\mathbb{C}^n$. We introduce some notations.

Let E be a subset of a metric space X. Denote by $\delta(E)$ the diameter of E and define for $p \geq 0$,

$$\delta^p(E) = [\delta(E)]^p \qquad p > 0$$

$$\delta^0(E) = \begin{cases} 1 & \text{if } E \neq \emptyset \\ 0 & \text{if } E = \emptyset . \end{cases}$$

Define

$$H^p_\varepsilon(E) = \inf\left\{\sum_1^\infty \delta^p(E_n) : E \subset \bigcup E_n, \quad \delta(E_n) < \varepsilon\right\}$$

$$H^p(E) = \lim_{\varepsilon \to 0} H^p_\varepsilon(E).$$

H^p is called the Hausdorff measure of dimension p. Note that $H^0(E)$ is the number of points of E.

We shall use the following particular case of a lemma due to Federer [4].

LEMMA 15 [14]. Let X be a metric space and f a map from X into $\mathbb{R}^n$. Suppose f is Lipschitz and set $\mathrm{Lip}(f) = \lambda$. Let $\alpha \geq 0$ $\beta > 0$ and suppose that $H^{\alpha+\beta}(X) < \infty$. Then

$$\int H^\alpha(f^{-1}(y))\, d^\beta y \leq \lambda^\beta H^{\alpha+\beta}(X).$$

For an elementary proof see [9].

We need also the following inequality due to Bishop ; for a proof see also [9].

Denote by $M_{n,k}$ the Grassmann manifold of complex k dimensional subspaces of $\mathbb{C}^n$. Let μ be the measure on $M_{n,k}$ invariant under the action of the unitary group and normalized by $\mu(M_{n,k}) = 1$.

LEMMA 16. *There exists a constant* a_n *(for* $n \geq 2$*) such that*

$$\int_{M_{n,n-k}} H^{\alpha}(Y \cap P)\, d\mu(P) \leq a_n\, H^{2k+\alpha}(Y)$$

for $\alpha \geq 0$, $0 < k < n$, *and* $Y \subset \{z \in \mathbb{C}^n \ ; \ \|z\| \geq 1\}$.

THEOREM 17. *Let* A *be a uniform algebra on a compact set* X *in* $\mathbb{C}^n$. *Suppose that* A *contains the polynomials and that the maximal ideal space of* A *is identified with* X. *Let* z_0 *be a point not in the Shilov boundary of order* $k-1$ *associated with* A. *Then* z_0 *has a neighborhood in* X *which is an analytic variety* V_z *of dimension* k *if and only if there exists a ball* $B(z_0,r)$ *such that*

$$H^{2k}(X \cap B(z_0,r)) < \infty.$$

Moreover the function of A *are analytic in* V_{z_0}.

Proof. Applying lemma 15 to the function $f(z) = \|z - z_0\|$, we see that for almost every r

$$H^{2k-1}(S(z_0,r)) < \infty.$$

Now apply lemma 16 with $\alpha = 0$ and $Y_r = S(z_0,r) \cap X$; observe that $H^{2k}(Y_r) = 0$ if $H^{2k-1}(Y_r) < \infty$. Thus we can choose an affine subspace P of dimension $n-k$ such that

$$(1) \qquad Y_r \cap P = \emptyset \quad \text{and} \quad z_0 \in P.$$

P is defined by k linear equations

$$\ell_1(z) = \alpha_1, \ldots \ell_k(z) = \alpha_k.$$

Define $F = (\ell_1, \ldots, \ell_k)$. Observe now that $X_{z_0,r} = X \cap B(z_0,r)$ is A-convex and that Σ^o_{k-1} the Shilov boundary of order $k-1$ relative to $A_{X_{z_0,r}}$ is contained in $S(z_0,r) \cap X$ (we suppose that $B(z_0,r) \cap \Sigma_{k-1} = \emptyset$). Consequently relation (1) implies that $F(z_0) \notin F(\Sigma^o_{k-1})$.

Clearly F is Lipschitz on X, so it follows from lemma 15 that

$$\int H^o(F^{-1}(y))\, d^{2k}y \leq \lambda^{2k} H^{2k}(B(z_0,r) \cap X).$$

Hence $F^{-1}(y)$ is finite for almost every $y \in \mathbb{C}^k$. By theorem 11 a neighborhood of z_0 has the structure of an analytic variety of dimension k.

On the other hand it is classical that the 2k-Hausdorff measure of an analytic variety of dimension k is locally finite see [10] for example.

Let Ω be a domain in $\mathbb{C}^n$ and let $F = (f_1, \ldots, f_p)$ be a p-tuple of analytic functions in Ω. We assume that the functions $(f_1, \ldots, f_p)$ separate points on Ω. Let K be a compact set in Ω and write A for the uniform algebra generated in $\mathscr{C}(K)$ by $(f_1, \ldots, f_p)$ and constants. The problem is to describe $\mathfrak{M}$ the maximal ideal space of A. Since $\mathfrak{M}$ is isomorphic with $h(\mathcal{K})$ the polynomially convex hull of $\mathcal{K} = F(K)$ in $\mathbb{C}^p$, the problem is equivalent to the description of $h(\mathcal{K})$.

THEOREM 18. Let $\mathcal{K} \subset \mathbb{C}^p$ be the image under a holomorphic map of a compact set in $\mathbb{C}^n$ as described above. Then there exists analytic structure of dimension n in an open dense set U of $h(\mathcal{K}) \setminus \Sigma_{n-1}$, where Σ_{n-1} denotes the Shilov boundary of order $n-1$ relative to the algebra of polynomials in $\mathbb{C}^p$. More precisely, each point p in U has a neighborhood in $h(\mathcal{K})$ which is a branched analytic cover of

<u>dimension</u> n.

For $G = (g_1, \ldots, g_n) \in A^n$ and $z \in \mathbb{C}^n$, put

$$G^{-1}(z) = \{m \in \mathfrak{M} / (g_1, \ldots, g_n)(m) = z\}.$$

We shall use the following theorem of Bishop, see [2] p. 484.

THEOREM 19. <u>Fix</u> $G \in A^n$. <u>Then for almost every</u> $z \in \mathbb{C}^n$, $G^{-1}(z)$ <u>is finite</u>.

This result appears in Bishop proof of the Oka theorem, for the case $n = 1$, see also [12].

Proof of theorem 18. Let x be a point in $h(\mathcal{K}) \setminus \Sigma_{n-1}$. Write $\mathcal{K}_{x,r}$ for the polynomially convex set $h(\mathcal{K}) \cap B(x,r)$. Suppose also that $B(x,r) \cap \Sigma_{n-1} = \emptyset$. Observe first that the Shilov boundaries of order $k \leq n-1$ relative to $\mathcal{K}_{x,r}$ and the algebra of polynomials are contained in $S_{x,r}$, as follows from the choice of r and the generalized maximum principle.

We shall prove the existence of an n-tuple $I = (i_1, \ldots, i_n)$ such that for any $r > 0$, the image of $\mathcal{K}_{x,r}$ under the projection $\pi_I = (z_{i_1}, \ldots z_{i_n})$ is of positive measure in $\mathbb{C}^n$. Denote by $|E|$ the Lebesgue measure of the set E in the appropriate $\mathbb{C}^k$.

There exists an index i_1 such that $|z_{i_1}(\mathcal{K}_{x,r})| > 0$. Suppose not ; then for every j, $1 \leq j \leq p$, $|z_j(\mathcal{K}_{x,r})| = 0$. Using the symbolic calculus and the fact that $H(K) = \mathcal{C}(K)$ for a compact set in the plane of measure zero, we see that the polynomials are dense in $\mathcal{C}(\mathcal{K}_{x,r})$ and so $x \in \Sigma_o \subset \Sigma_{n-1}$, and this contradicts $x \notin \Sigma_{n-1}$.

Suppose we have proved the existence of $J = (i_1, \ldots, i_j)$ $j \leq n-1$ such that the image of $\mathcal{K}_{x,r/2}$ under $\pi_J = (z_{i_1}, \ldots z_{i_j})$ is of positive measure in $\mathbb{C}^j$; we then construct $J' = (i_1, \ldots, i_j, i_{j+1})$ with the same property in $\mathbb{C}^{j+1}$. Define the projection $\pi_{J,\ell} = (z_{i_1}, \ldots z_{i_j}, z_\ell)$. Suppose that for every $1 \le \ell \le p$ $|\pi_{J,\ell}(\mathcal{K}_{x,r})| = 0$. Then for almost all $\alpha = (\alpha_1, \ldots \alpha_j) \in \mathbb{C}^j$ the α-section of $\pi_{J,\ell}(\mathcal{K}_{x,r})$ is of measure zero in $\mathbb{C}$. As before we see that the algebra of polynomials restricted to $Z_\alpha = \pi_J^{-1}(\alpha) \cap \mathcal{K}_{x,r}$ is dense in $\mathcal{C}(Z_\alpha)$ and, as we have already seen, Z_α is contained in $\Sigma_j \subset S(x,r)$. It follows that $\pi_J(\mathcal{K}_{x,r/2})$ is contained in a set of measure zero, which contradicts the choice of J.

Thus we have proved the existence of π_I such that for every $r > 0$ $|\pi_I(\mathcal{K}_{x,r})| > 0$. Using theorem 19, we see that for almost every $z \in \mathbb{C}^n$ $\pi_I^{-1}(z) \cap h(\mathcal{K})$ is finite. Therefore there exists $x' \in \mathcal{K}_{x,r}$ and a ball $B(x',r')$ such that $\pi_I(x') \notin \pi_I(S(x',r') \cap h(\mathcal{K}))$; it is sufficient to separate x' from other points of $\pi_I^{-1}(\pi_I(x'))$. At this point we used that $|\pi_I(\mathcal{K}_{x,r})| > 0$.

Since the boundary Σ'_{n-1} relative to the algebra of polynomials restricted to $\mathcal{K}_{x',r'}$ is contained in $S(x',r')$, we can apply theorem 11 from which it follows that a neighborhood of x' in $h(\mathcal{K})$ is an analytic set of dimension n.

Example. We use an example of Wermer [13] to show that $h(\mathcal{K}) \setminus \Sigma_{n-1}$ can contain something more than $\mathcal{K}$.

Let B_R denote the bycilinder in $\mathbb{C}^2$ of radius R,

$$B_R = \{(z,w)/\ |z| \le R,\ |w| \le R\}.$$

Let F be the map from B_R into $\mathbb{C}^3$ defined as follows

$$F(z,w) = (z\ ,\ zw\ ,\ zw^2 - w).$$

Write $(u\ ,\ v\ ,\ t)$ for coordinates in $\mathbb{C}^3$. Then for every R, $K_R = F(B_R)$ is contained in the hypersurface S defined by

$$S = \{(u,v,t) \in \mathbb{C}^3 / v(v-1) = tu\}.$$

It is clear that $h(K_R)$ is also contained in S. J. Wermer proved in [13] that

$h(K_1)$ is the union of K_1 and the open disc $\{(z,0,1)/ |z| < 1\}$. Here we assert that for R large enough $h(K_R) \setminus K_R$ contains an open set of S, and clearly this set cannot be in Σ_1 by the maximum principle on varieties. If we consider the uniform algebra A on B_R generated by the functions (z, zw, zw^2-w) the spectrum $\mathfrak{M}_A$ of A is isomorphic with $h(K_R)$ and then $\mathfrak{M}_A \setminus \Sigma_1$ contains a non trivial analytic set of dimension two.

Proof of the assertion. Observe first that the projection of K_R on the t-axis is a disc centered at the origin. In addition if R is large enough we can suppose that it contains the unit disc D. Fix $t_o \in D \setminus \{0\}$. The section of K by the hyperplane $t = t_o$ can be parametrized by

$$w \to \left(\frac{t_o+w}{w^2}, \frac{t_o+w}{w}, t_o\right)$$

with w varying in $D_{t_o} = \left\{w/ |w| \leq R \quad \left|\frac{t_o + w}{w^2}\right| \leq R\right\}$.

If $R > \sqrt{2}$, then D_{t_o} contains the circle of radius R. Also $0 \notin D_{t_o}$. Let P be a polynomial in $\mathbb{C}^3$ such that $\|P\|_{K_R} \leq 1$. In particular

$$\left|P\left(\frac{t_o+w}{w^2}, \frac{t_o+w}{w}, t_o\right)\right| \leq 1 \quad \text{for} \quad |w| = R,$$

and so the same inequality is true for $|w| > R$ considered as an open set in the Riemann sphere. Then $h(K_R)$ contains the image of $(D \setminus \{0\}) \times (|w| > R)$ under the map ψ defined by

$$\psi(t,w) = \left(\frac{t+w}{w^2}, \frac{t+w}{w}, t\right).$$

Observe that the image of ψ is contained in S, and that ψ is an immersion. Then the image is an open set of S containing the punctured disc

$$D' = \{(0,1,t) \quad |t| < 1 \quad t \neq 0\},$$

and it is clear that $D' \cap K_R = \emptyset$. This proves the assertion.

Question. Under the hypotheses of theorem 18, is it true that every point of $\mathfrak{M}_A \setminus \Sigma_{n-1}$ has a neighborhood in $\mathfrak{M}_A$ which is an analytic space of dimension n.

5. A CONDITION FOR ANALYTIC STRUCTURE IN $\mathfrak{M}_A$.

Let A be a uniform algebra on a compact space X. The spectrum $\mathfrak{M}$ is a subset of the dual A', and it can be given two topologies : the weak one induced by $\sigma(A',A)$ and the topology induced by the dual norm. We shall consider the Hausdorff measure H^p on $\mathfrak{M}$ associated with the metric induced by the dual norm. Let x be a point in $\mathfrak{M}$. If $\| \ \|$ denote the dual norm we write $U(x,r)$ for the set

$$\{y \in \mathfrak{M},\ \|x-y\| \leq r\}.$$

We can then formulate a condition for the existence of analytic structure in $\mathfrak{M}$.

THEOREM 20. Let x be a point in $\mathfrak{M} \setminus \Sigma_{n-1}$, $n \geq 0$. The following conditions are equivalent

i) There exists a weak neighborhood U_x of x, which is a branched analytic cover of dimension n such that for every $f \in A$, $f|_{U_x}$ is holomorphic in U_x.

ii) The norm topology and the weak topology coincide in a neighborhood of x and there exists $r > 0$ such that $H^{2n}(U(x,r)) < \infty$.

Proof. ii) $\Rightarrow$ i). We prove firstly that for every $F = (f_1, \ldots, f_n) \in A^n$ $F^{-1}(z) \cap U(x,r)$ is finite for almost every $z \in \mathbb{C}^n$. In fact the function F is Lipschitz with respect to the metric induced by the norm of A'. Then by lemma 15, there exists $C > 0$ such that

$$\int H^0\left[F^{-1}(z) \cap U(x,r)\right] d^{2n}z \leq C\, H^{2n}(U(x,r)).$$

It follows that $H^0\left[F^{-1}(z) \cap U(x,r)\right]$ is finite for almost every z.

Since the weak topology and the norm topology coincide in a neighborhood of x, there exist N function $f_1, \dots, f_N \in A$ such that if

$$\varphi(y) = \sup_{i \leq N} |f_i|(y)$$

then $\varphi(x) = 0$ and the weak neighborhood $\{y \in \mathfrak{M} \ \varphi(y) < 1\}$ is contained in $U(x,r)$. Using that φ is Lipschitz with respect to $\| \ \|$, we have

$$\int_0^1 H^{2n-1}(\varphi^{-1}(t))\, dt \leq C\, H^{2n}(U(x,r)).$$

It follows that for almost every $0 < t < 1$, $H^{2n-1}(\varphi^{-1}(t)) < \infty$. Let t_o be such that $Z_o = \{y \in \mathfrak{M} / \varphi(y) = t_o\}$ is of H^{2n}-measure zero. Let K be the image of Z_o in $\mathbb{C}^N$ under the map

$$\Phi = (f_1, \dots, f_N).$$

It is clear that $H_\bullet^{2n}(K) = 0$ ($H_\bullet^{2n}$ denote here the 2n-Hausdorff measure in $\mathbb{C}^N$) and so by lemma 16 there exists a subspace P of dimension $N - n$ such that $P \cap K = \emptyset$. Using the equations of P, we construct $G = (g_1, \dots, g_n)$, $G \in A^n$, such that $G(x) = 0$ and $G^{-1}(0) \cap Z_o = \emptyset$.

Consider now the restriction of A to the A-convex neighborhood Z of x defined by

$$Z = \{y \in \mathfrak{M},\ \varphi(y) \leq t_o\}.$$

By the local maximum principle the Shilov boundary of order $(n-1)$ of $A|_Z$ is contained in Z_o Using the first part of the proof we see that the hypotheses of theorem 11 are satisfied by the function G constructed above, for the algebra A_Z.

The fact that i) $\Rightarrow$ ii) is a direct consequence of the Montel theorem as stated in [6] p. 234, and the fact that analytic sets of dimension n are locally of H^{2n} finite measure.

[1] BASENER, R. F. A generalized Shilov boundary and analytic structure. Proc. Amer. Math. Soc. 47 (1975), 98-104.

[2] BISHOP, E. Holomorphic completions, analytic continuation, and the interpolation of semi-norms. Ann. Math. 78 (1963), 468-500.

[3] CARTAN, H. Calcul différentiel. Paris, Hermann (1967).

[4] FEDERER, H. Some integral geometric theorems. Trans. Amer. Math. Soc. 77 (1954), 238-261.

[5] GLEASON, A. M. The abstract theorem of Cauchy-Weil. Pacific J. Math. 12 (1962), 511-525.

[6] GUNNING, R. C. and ROSSI, H. Analytic functions of several complex variables. Prentice Hall, Englewood Cliffs, N. J. (1965).

[7] GRAUERT, H. and REMMERT, K. Komplexe Räume. Math. Ann. 136 (1958), 245-318.

[8] ROSSI, H. Holomorphically convex sets in several complex variables. Ann. Math. 74, (1961), 470-493.

[9] SHIFFMANN, B. On the removal of singularities of analytic sets. Michigan Math. J. 15 (1968), 110-120.

[10] STOLZENBERG, G. Volume, limits and extensions of analytic varieties. Lecture Notes in Math. 19. Berlin, Springer-Verlag (1966).

[11] STOUT, E. L. The theory of uniform algebras. Bogden Quigley, N. Y. (1971).

[12] WERMER, J. Banach algebras and several complex variables. Markham, Chicago (1971)

[13] WERMER, J. An example concerning polynomial convexity. Math. Ann. 139 (1959), 147-150.

BOUNDARY VALUES FOR THE SOLUTIONS OF THE $\bar{\partial}$-EQUATION AND APPLICATION TO THE NEVANLINNA CLASS

H. Skoda

1) Introduction and notations.

Let Ω be a bounded, strictly pseudoconvex, open set in $\mathbb{C}^n$, of class C^2, that is:

$$\Omega = \{z | \rho(z) < 0\},$$

where ρ is a function of class C^2, defined in a neighbourhood of $\bar{\Omega}$, strictly plurisubharmonic, and verifying $d\rho \neq 0$ in a neighbourhood of $\partial\Omega$.
$\delta(z)$ is the boundary distance.
We are willing to resolve in Ω the equation

$$(1) \quad d''U = f,$$

where f is a $(0,1)$ form in Ω, d''-closed, and U a function, whose boundary value $u = U_{|\partial\Omega}$ is, in some sense, in $L^p(\partial\Omega)$ for some p, $1 \leq p \leq +\infty$.
The problem is to give good, sufficient, conditions about f, such that there exists a solution U of (1) with good boundary value.
We shall essentially consider the two following cases:

- $f \in L^1(\Omega)$, or the coefficients of f are bounded measures over Ω.
- $f_{|\partial\Omega} \in L^1(\partial\Omega)$ in some sense.

To simplify, we limit our study to $(0,1)$ forms, but similar results are true when f is a (p,q) form and U a $(p,q-1)$ form.
The motivations for this problem are the following:
The theory of H^p spaces in Ω naturally requires such estimations for the d''-equation. I shall particularly give two examples:

The first example is the "Corona problem" in its concret and simplest form:

let $g_1, g_2 \in H^\infty(\Omega)$ with $|g_1| + |g_2| \geq c > 0$

over Ω (where c is a constant), does there exist h_1 and $h_2 \in H^\infty(\Omega)$ such that $g_1h_1 + g_2h_2 = 1$ in Ω. By standard methods of [1] and [7], it is sufficient to solve an equation $d''U = f$, where f has for coefficients bounded measures on Ω "of Carleson type", and if we are able to obtain a solution U with boundary value in $L^\infty(\partial\Omega)$, then we are able to resolve the "Corona problem".

The second example is the problem to characterize the zeros of functions in $H^p(\Omega)$.
Let X be an hypersurface in Ω. We are willing to find a holomorphic function F such that $X = \{z \mid F(z) = 0\}$ and that F has a good boundary value. If X verifies the Blaschke condition:

$$(2) \int_X \delta(z)d\sigma(z) < +\infty,$$

where $d\sigma$ is the area element of X, then by classical arguments (cf. for example [4]), it is sufficient to solve our equation $d''U = f$, where the coefficients of f are bounded measures on Ω, and we obtain F by:

$$\mathrm{Log}|F| = \mathrm{Re}\ U.$$

Therefore a good solution of (1) will give a function F in the Nevanlinna class or in $H^\infty(\Omega)$.

In this paper, we shall not obtain the solution of the "Corona problem", but we shall prove that the Blaschke condition is characteristic of zeros of functions in the Nevanlinna class.

The results are announced in [23] and [24], and the detailed proof will appear in [25]. Similar results were announced recently by G.M. Henkin [6].

2) Generalized solutions of the d"-equation.

We shall study the problem of the d''-equation in a slightly new form, to insist on the boundary value of the solution. We notice that if φ is a $(n,n-1)$ form of C^∞ class in $\overline{\Omega}$, and d''-closed, and if U is of class C^1 in $\overline{\Omega}$, then the equation $d''U = f$ implies:

$$\int_\Omega f \wedge \varphi = \int_\Omega d''U \wedge \varphi = \int_\Omega d(U \wedge \varphi) = \int_{\partial\Omega} u \wedge \varphi.$$

That is :

$$(3) \quad \int_{\Omega} f \wedge \varphi = \int_{\partial\Omega} u\varphi ,$$

for all $\varphi \in C^1_{n,n-1}(\overline{\Omega})$ such that $d''\varphi = 0$, and with $u = U_{|\partial\Omega}$.

We shall directly solve the integral relation (3) which keeps a signification when f has measure coefficients in Ω and when u is a function defined only over $\partial\Omega$. When u verifies (3), we shall say that u is a tangential solution for the d''-equation, and we shall note abusively (3):

$$d''_b u = f,$$

because when $f \in C^1_{0,1}(\overline{\Omega})$, then we shall obtain the tangential $\overline{\partial}_b$ operator of J. Kohn over $\partial\Omega$ (cf. [2]).

When we obtain u solution of (3), we build U solution of (1) in Ω by a standard process, and we have:

<u>Proposition 1</u>. a) If f is in $L^1_{0,1}(\Omega)$ (or f has bounded measure coefficients), and if $u \in L^1(\partial\Omega)$ verifies (3), then there exists $U \in L^1(\Omega)$ such that:

$$d''U = f \quad , \quad \text{in } \Omega,$$

and such that for all $\varphi \in C^1_{n,n-1}(\overline{\Omega})$ we have:

$$-\int_{\Omega} U \wedge d''\varphi = \int_{\Omega} f \wedge \varphi - \int_{\partial\Omega} u \wedge \varphi,$$

(this means that U has boundary value u in the sense of the Stokes formula).

b) If $f \in C^0_{0,1}(\overline{\Omega})$, and if $u \in C^0(\partial\Omega)$, then there exists $U \in C^0(\overline{\Omega})$ such that $d''U = f$ in Ω, and $u = U_{|\partial\Omega}$.

U is given by the following formula:

$$U(z) = \int_{\Omega} K_M(z,\zeta) \wedge f(\zeta) + \int_{\partial\Omega} K_M(z,\zeta)\, u(\zeta),$$

where $K_M(z,\zeta)$ is the Bochner-Cauchy-Martinelli kernel:

$$K_M(z,\zeta)^1 = c_n \sum_{i=1}^{n} (-1)^{i-1} \frac{z_i-\zeta_i}{|z-\zeta|^{2n}} \left(\bigwedge_{j\neq i} d\bar{\zeta}_j \right) \wedge \omega(\zeta),$$

where $\omega(\zeta) = \bigwedge_{j=1}^{n} d\zeta_j$, and c_n is a constant.

The mean interest of the formula (3) will be to give for the tangential solution u a simpler formula connected with the strict pseudoconvexity of Ω and without the Cauchy-Martinelli term.

3) Results for the d"-equation when $f \in L^1(\Omega)$.

If f is a (p,q) form in Ω, with bounded measures coefficients, which is canonically written:

$$f = \sum_{I,J} f_{I,J}\, dz_I \wedge d\bar{z}_J \quad ,$$

$|I| = p$, $|J| = q$, we put:

$$\|f\|_1 = \sum_{I,J} \|f_{I,J}\|_1 \quad ,$$

where $\|f_{I,J}\|_1$ is the integral over Ω of the positive measure $|f_{I,J}|$.

For each $x \in \partial\Omega$, we call T_x^c the complex tangential hyperplane at x and we call $A(x,t)$ the Hörmander-Koranyi ball, centered in x and of radius $t > 0$:

$$A(x,t) = \{z \in \Omega | \exists\ \zeta \in T_x^c,\ |\zeta-z| < t,\ |\zeta-x| < t^{\frac{1}{2}}\}$$

cf. [8].
dS is the volume element on $\partial\Omega$.

We have the following result:

Theorem 1. If f is a $(0,1)$ form on Ω, d"-closed and if the coefficients of f and of the form $\delta^{-\frac{1}{2}}\, d''\rho \wedge f$ are bounded measures on Ω, then there exists $u \in L^1(\partial\Omega)$ such that:

$$d_b'' u = f \ ,$$

in the sense of (3), and:

$$\int_{\partial\Omega} |u| \, ds \leq C(\Omega) \left(\|f\|_1 + \|\delta^{-\frac{1}{2}} d''\rho \wedge f\|_1 \right) ,$$

where $C(\Omega)$ is a constant which depends only of Ω.

The condition about $d''\rho \wedge f$ is in fact a restriction concerning the coefficients of f in the tangential directions. These coefficients must not be too large.
There exists a counterexample, which shows that the theorem is false if we drop the condition about $d''\rho \wedge f$.
If the coefficients of f are regular enough, we obtain u in $L^\infty(\partial\Omega)$:

<u>Theorem 1'</u>. If the measure μ defined by $\mu = |f| + |\delta^{-\frac{1}{2}} d''\rho \wedge f|$ verifies the following condition:

$$\exists\, C_1 > 0 \text{ and } \alpha > n, \text{ such that } \forall\, x \in \partial\Omega, \forall\, t > 0$$
$$\mu[A(x,t)] \leq C_1 t^\alpha,$$

then the solution u, given by the theorem 1, is in $L^\infty(\partial\Omega)$ and we have:

$$\|u\|_{L^\infty(\partial\Omega)} \leq C(\Omega,\alpha) \sup_{x\in\partial\Omega, t>0} t^{-\alpha} \mu[A(x,t)] .$$

The measures which verify the hypothesis of the theorem 1' with $\alpha = n$, were introduced by Hörmander [8].
We have then existence in $L^\infty(\partial\Omega)$ for a measure μ scarcely more regular than a Hörmander's measure. I conjecture that the theorem is true with $\alpha = n$ and that it is the first fundamental step for the solution of the "Corona problem".
The solution u is given by an implicit formula, which is particularly simple in the case where Ω is strictly convex. In this case, we put:

$$P_j(\zeta) = \frac{\partial\rho}{\partial\zeta_j}(\zeta) \quad \text{and} \quad Q_j(z) = P_j(z)$$

and we have when $z \in \partial\Omega$:

$$u(z) = c_n \int_{\Omega_\zeta} \nu(z,\zeta) \wedge f(\zeta) ,$$

where $\nu(z,\zeta)$ is the differential form:

$$\nu(z,\zeta) = \sum_{i=1}^{n} \frac{(-1)^{i}\,\rho(\zeta)\,Q_i}{[-\rho+\langle P,\zeta-z\rangle]^n \langle Q,\zeta-z\rangle} \bigwedge_{j\neq i} d''_{\zeta}P_j \wedge \omega(\zeta)$$
$$+ \sum_{i<j} \frac{(-1)^{i+j}\,(P_iQ_j-P_jQ_i)}{[-\rho+\langle P,\zeta-z\rangle]^n \langle Q,\zeta-z\rangle}\, d''\rho \wedge \bigwedge_{k\neq i,j} d''P_k \wedge \omega(\zeta),$$

where ρ is a function of ζ and $\langle \zeta,z\rangle = \sum_{i=1}^{n} \zeta_i z_i$.

The second term gives the condition over $d''\rho \wedge f$.
We have another expression for ν:

$$\nu(z,\zeta) = c'_n \frac{\rho^n\, d'_{\zeta}(\langle Q,\zeta\rangle) \wedge (d'd''\mathrm{Log}-\rho)^{n-1}}{[-\rho+\langle P,\zeta-z\rangle]^n \langle Q,\zeta-z\rangle} .$$

We shall later give much information about the proof.

4) Results for the d''-equation when $f \in L^1(\partial\Omega)$.

We consider now the case where f has a "boundary value". We suppose for the moment that $f \in C^1_{0,1}(\overline{\Omega})$. For $x \in \partial\Omega$, let $f_b(x)$ the restriction of the $\mathbb{C}$-antilinear form $f(x)$ to the complex tangent hyperplane T^c_x , we set:

$$\|f_b(x)\| = \sup_{\substack{\tau\in T^c_x \\ \|\tau\|=1}} |\langle f(x),\tau\rangle| ,$$

and:

$$\|f_b\|_{L^p(\partial\Omega)} = \left[\int_{\partial\Omega} \|f_b(x)\|^p \, dS(x) \right]^{\frac{1}{p}} .$$

Theorem 2. If f is a d''-closed $(0,1)$ form of class C^1 in $\overline{\Omega}$, there exists a solution u of $d''_b u = f$, in the sense of (3), which is given by an integral kernel K:

$$(\underline{K}f)(z) = u(z) = \int_{\partial\Omega} K(z,\zeta) \wedge f(\zeta),$$

where $z \in \partial\Omega$. The operator $\underline{K}$ is continuous from $L^r(\partial\Omega)$ in $L^s(\partial\Omega)$:

$$\|\underline{K}f\|_{L^s} \leq C(\Omega,r,s)\, \|f_b\|_{L^r} ,$$

with $1 \le r, s \le +\infty$ and $\frac{1}{s} > \frac{1}{r} - \frac{1}{2n}$.

This result is similar to the result of Folland and Stein [3], but we obtain the values $r = 1$ and $r = +\infty$.

Naturally, the theorem is valid even if f is not of class C^1 in $\bar{\Omega}$. For example, it is sufficient to suppose that f is a limit of a sequence $f_n \in C^1_{0,1}(\bar{\Omega})$ such that $f_n \to f$ in $L^1(\Omega)$ and such that $\|f_{b,n}\|_{L^r(\partial\Omega)}$ is a bounded sequence.

When Ω is strictly convex, the kernel K is given by:

$$K(z,\zeta) = c_n \sum_{i<j} \frac{(-1)^{i+j-1}(P_iQ_j - P_jQ_i)}{\langle P,\zeta-z\rangle^{n-1}\langle Q,\zeta-z\rangle} \bigwedge_{k\neq i,j} d''_\zeta P_k \wedge \omega(\zeta) .$$

Another expression for K is the following:

$$K(z,\zeta) = c'_n \frac{d'_\zeta(\langle Q,\zeta\rangle) \wedge d'\rho \wedge (d'd''\rho)^{n-2}}{\langle P,\zeta-z\rangle^{n-1}\langle Q,\zeta-z\rangle} .$$

5) Summary of the proof of theorem 1.

We use methods which are similar to the methods of G. Henkin [5], N. Kerzman [9], I. Lieb [20] and N. Øvrelid [22].

Let μ be the Cauchy-Leray differential form:

$$\mu = \langle\xi,\zeta-z\rangle^{-n} \sum_{i=1}^{n} (-1)^{i-1} \xi_i \, (\bigwedge_{j\neq i} d\xi_j) \wedge \omega(\zeta-z).$$

μ is defined over the set E defined by:

$$E = \{(\xi,\zeta,z) \in \mathbb{C}^{3n} \mid \langle\xi,\zeta-z\rangle \neq 0\}.$$

Let $\pi : E \to \mathbb{C}^{2n} \setminus \Delta$, $(\xi,\zeta,z) \to (\zeta,z)$, the canonical projection, where Δ is the diagonal of $\mathbb{C}^n \times \mathbb{C}^n$.

μ is closed on E : $d\mu = 0$.

Let s_b the Bochner-Cauchy-Martinelli section of π :

$$s_b : \mathbb{C}^n \times \mathbb{C}^n \setminus \Delta \to E ,$$

$$s_b(\zeta,z) = (\bar{\zeta}-\bar{z},\zeta,z).$$

Then we have classically:

$$d(s_b^*\mu) = d''(s_b^*\mu) = c_n[\Delta],$$

in the sense of currents in $\mathbb{C}^n \times \mathbb{C}^n$, where $[\Delta]$ is the current of integration over Δ.
Therefore, we have with the notations of §1 :

$$c_n \int_\Omega f \wedge \varphi = c_n \int_{\Delta\cap(\Omega\times\Omega)} f(\zeta) \wedge \varphi(z) = \int_{\partial(\Omega\times\Omega)} s_b^*\mu \wedge f(\zeta) \wedge \varphi(z).$$

Let s_h an another section of E, defined on $\partial(\Omega\times\Omega) \smallsetminus \Delta$.
As s_b and s_h are homotopic and as the form $\mu \wedge f(\zeta) \wedge \varphi(z)$ is closed on $\pi^{-1}(\bar\Omega\times\bar\Omega\smallsetminus\Delta)$, we have by Stokes formula:

$$c_n \int_\Omega f \wedge \varphi = \int_{\partial(\Omega\times\Omega)} s_h^*\mu \wedge f(\zeta) \wedge \varphi(z).$$

As $\partial(\Omega\times\Omega) = \partial\Omega_s \times \Omega_z + \Omega_s \times \partial\Omega_z$, we choose s_h such that s_h is <u>holomorphic in z</u> when $\zeta \in \partial\Omega$.
Then by bidegree considerations, we have:

$$\int_{\partial\Omega_\zeta\times\Omega_z} s_h^*\mu \wedge f(\zeta) \wedge \varphi(z) = 0.$$

Then we have:

$$c_n \int_\Omega f \wedge \varphi = \int_{\Omega_\zeta\times\partial\Omega_z} s_h^*\mu \wedge f(\zeta) \wedge \varphi(z).$$

That is we have a tangential solution u of the d''-equation in the sense of (3), which is given by:

$$u(z) = \frac{1}{c_n} \int_{\Omega_\zeta} s_h^*\mu \wedge f(\zeta),$$

where $s_h^*\mu$ is in fact the component of bidegree $(n,n-1)$ in ζ and $(0,0)$ in z of $s_h^*\mu$.
To obtain u, it is therefore sufficient to build a nice section s_h on $\partial(\Omega\times\Omega) \smallsetminus \Delta$.
In the case where Ω is strictly convex, we take:

$$\xi_h(\zeta,z) = \frac{-\rho(\zeta)}{\langle Q,\zeta-z\rangle} Q(z) + P(\zeta).$$

The announced formula is a consequence of a computation, because then ρ and $d''\rho$ appear in $s_h^*\mu$.

In the general case, we take for P a more complicated function $P(\zeta,z)$ of Lieb-Ramirez ([20]), which is holomorphic in z, and we take:

$$Q(\zeta,z) = P(z,\zeta)$$

and

$$\xi_h = - \frac{\rho}{\langle Q,\zeta-z\rangle} Q + P .$$

In the case of the ball, the section s_h is closely connected with the Poisson-Szegö representation formula for holomorphic functions.

To obtain the theorem 2, we consider now a discontinuous section of E over $\partial(\Omega\times\Omega) \setminus \Delta$ with a discontinuity over $\partial\Omega \times \partial\Omega$. We take:

$$\xi_h(\zeta,z) = P(\zeta,z) , \quad \text{when } \zeta \in \partial\Omega \text{ and } z \in \Omega,$$

$$\xi_h(\zeta,z) = Q(\zeta,z) , \quad \text{when } z \in \partial\Omega \text{ and } \zeta \in \Omega.$$

We consider the chain of homotopy between s_b and s_h:

$$F_1(t,\zeta,z) = t\,\xi_b + (1-t)P, \quad \zeta \in \partial\Omega,\ z \in \Omega,\ t \in [0,1]$$

$$F_2(t,\zeta,z) = t\,\xi_b + (1-t)Q, \quad z \in \partial\Omega,\ \zeta \in \Omega,\ t \in [0,1].$$

Then Stokes formula gives:

$$\int_\Omega s_b^*\mu \wedge f(\zeta) \wedge \varphi(z) = \int_{I\times\partial\Omega\times\partial\Omega} (F_1^*\mu - F_2^*\mu) \wedge f(\zeta) \wedge \varphi(z) ,$$

where $I = [0,1]$. Therefore we have:

$$c_n \int_\Omega f \wedge \varphi = \int_{I\times\partial\Omega\times\partial\Omega} (F_1^*\mu - F_2^*\mu) \wedge f(\zeta) \wedge \varphi(z),$$

and we obtain a solution u of (3) given by:

$$u(z) = \frac{1}{c_n} \int_{I\times\partial\Omega_\zeta} (F_1^*\mu - F_2^*\mu) \wedge f(\zeta),$$

(where we take only the component in $F_j^*\mu$ of degree 1 in t, $(n,n-2)$ in ζ and $(0,0)$ in z).
The announced formula is a consequence of a computation.
The solutions u of theorem 1 and 2 are closely connected but different.

6) <u>Results for the d'd"-equation and the Nevanlinna class.</u>

Using the (n,n) form $(\frac{i}{2})^n \bigwedge_{j=1}^{n} (dz_j \wedge d\bar{z}_j)$, we identify 0-current and (n,n) current.

Let θ be a positive, closed, current of bidegree (1,1) in Ω, which is canoncially written:

$$\theta = i \sum_{j,k} \theta_{j_k} \, dz_j \wedge d\bar{z}_k .$$

The positivity means that for all $\lambda \in \mathbb{C}^n$, the measure $\sum_{j,k} \theta_{j_k} \lambda_j \bar{\lambda}_k$ is positive.

The measure $\sigma = 2 \sum_{j=1}^{n} \theta_{jj}$ is called the trace-measure of θ.

We shall say that θ verifies the Blaschke condition if:

$$\int_\Omega \delta(z) \, d\sigma(z) < +\infty .$$

The most important case is the case of the current of integration on a complex hypersurface X of Ω:

$$\theta = [X] .$$

In this case, the measure $d\sigma$ is the area element on X, cf. P. Lelong [18] and [19].

We have the following result, where $\partial\Omega_\varepsilon$ is the real hypersurface $\rho(z) = -\varepsilon$, $\varepsilon > 0$, and dS_ε the Lebesgue measure on $\partial\Omega_\varepsilon$.

<u>Theorem 3</u>. If θ verifies the Blaschke condition and if the cohomology class of θ in $H^2(\Omega,\mathbb{C})$ is 0, there exists a plurisubharmonic function V in Ω such that:

$$i \, d'd''V = \theta ,$$

and

$$\sup_{\varepsilon>0} \int_{\partial\Omega_\varepsilon} V^+ \, dS_\varepsilon < +\infty ,$$

where $V^+ = \text{Sup}(V,0)$.

When $\theta = [X]$ we obtain the following corollary:

Theorem 4.

If X verifies the Blaschke condition, if the cohomology class of X in $H^2(\Omega,\mathbb{Z})$ is 0, and if $H^1(\Omega,\mathbb{R}) = 0$, then there exists a holomorphic function F in the Nevanlinna class such that:

$$X = \{z \in \Omega \mid F(z) = 0\},$$

and

$$\sup_{\varepsilon>0} \int_{\partial\Omega_\varepsilon} \operatorname{Log}^+|F| \, dS_\varepsilon < +\infty .$$

Several partial results were obtained before by G. Laville [16] and L. Gruman [4].
It is classical that Theorem 3 implies Theorem 4 (cf. P. Lelong [18]).
We shall prove the Theorem 3, using Theorem 1 about the d''-operator.
By means of a regularization and a passage to weak limit, it is not a restriction to suppose θ of class C^∞ in Ω. First, we shall prove that the Blaschke condition implies very strong properties of the coefficients of θ.
Let ξ_1 a field of unitary vectors over Ω.
Let ξ_2 another field of unitary vectors over Ω, such that for all $z \in \Omega$, $\xi_2(z)$ belongs to $T_z^c = \operatorname{Ker} d'\rho(z)$.
In the following we consider θ_z as a hermitian positive form on $\mathbb{C}^n$.

Lemma 1. There exist constants $C_1(\Omega)$ and $C_2(\Omega)$ such that:

a) $\int_\Omega \theta_z(\xi_2,\xi_2)\, d\lambda(z) \leq C_1(\Omega) \int_\Omega \delta(z)\, d\sigma(z)$,

it is the "Malliavin condition" (cf. [21])

b) $\int_\Omega |\theta_z(\xi_1,\xi_2)|\,[\delta(z)]^{\frac{1}{2}}\, d\lambda(z) \leq C_2(\Omega) \int_\Omega \delta(z)\, d\sigma(z)$,

we call it the "mixed condition".
($d\lambda$ is the Lebesgue measure).

We give a quick proof. By Stokes formula, we have:

$$\int_\Omega -\rho\,(i\, d'd''\rho)^{n-1} \wedge \theta = \int_\Omega i\, d'\rho \wedge d''\rho \wedge (i\, d'd''\rho)^{n-2} \wedge \theta .$$

The first integral is in fact the Blaschke integral and the second integral is in fact equivalent to:

$$\int_\Omega |d'_e|^2 \text{ Trace } \theta_{z|T_z^c} \, d\lambda(z).$$

To prove the inequality b), we simply use the Schwartz-inequality, because of the positivity of θ:

$$2[\delta(z)]^{\frac{1}{2}} |\theta_z(\xi_1,\xi_2)| \leq \delta(z) \theta_z(\xi_1,\xi_1) + \theta_z(\xi_2,\xi_2).$$

To simplify, we suppose now Ω strictly convex. To resolve the equation $i\,d'd''V = \theta$, we solve classically the equation:

$$i\,d\omega = \theta.$$

We decompose ω:

$$\omega = -\omega_1 + \omega_2 ,$$

where ω_1 and ω_2 are the components of bidegree (1,0) and (0,1) of ω.

ω_1 and ω_2 are explicitely given by:

$$\omega_2 = \sum_{j,k} [\int_0^1 tz_k \, \theta_{kj}(tz)dt] \, d\bar{z}_j$$

$$\omega_1 = \overline{\omega}_2 .$$

ω_2 is d''-closed. Let U a solution of:

$$d''U = \omega_2 .$$

Then an immediate computation shows that $V = 2 \text{ Re } U$ is a solution of:

$$i\,d'd''V = \theta.$$

It is therefore sufficient to solve:

$$d''U = \omega_2 ,$$

with a boundary value in $L^1(\partial\Omega)$ for U. Using theorem 1, it is therefore sufficient to prove that:

$$\int_\Omega |\omega_2| < +\infty$$

and

$$\int_\Omega \delta^{-\frac{1}{2}} |\, d''\rho \wedge \omega_2 \,| < +\infty .$$

In order to estimate ω_2, we need the following elementary lemma.

Lemma 2. Let $\alpha > -1$ and let g a measurable function. Then we have:

$$\int_\Omega [\delta(z)]^\alpha \left|\int_0^1 tg(tz)dt\right| d\lambda(z) \leq C(\Omega,\alpha) \cdot \int_\Omega [\delta(z)]^{\alpha+1} |g(z)| \cdot |z|^{-2n-1} d\lambda(z) .$$

Using the Lemma 2 with $\alpha = 0$, we obtain that the Blaschke condition implies that:

$$\int_\Omega |\omega_2| < +\infty .$$

Using now the Lemma 2 with $\alpha = -\frac{1}{2}$, we see that the "mixed condition" of Lemma 1 implies that:

$$\int_\Omega \delta^{-\frac{1}{2}} |d''\rho \wedge \omega_2| < +\infty .$$

In Lemma 1, we take $\xi_1 = \frac{z}{|z|}$ and we take for ξ_2 different tangential vector fields which form an orthonormal local frame of T_z^c for all $z \in \Omega$. We notice that the coefficient of $d''\rho \wedge \omega_2$ are precisely the "tangential" coefficients of ω_2.

Remark. Let $u = U_{|\partial\Omega}$ be the boundary value of the solution U of the equation:

$$d''U = \omega_2 ,$$

in the sense of (3). Then it is not very difficult to prove the following formula:

$$V(z) = \int_{\Omega_\zeta} - 2\, G(z,\zeta)\, d\sigma(\zeta) + \int_{\partial\Omega_\zeta} R(z,\zeta)\, 2 \operatorname{Re} u(\zeta) ,$$

where G is the Green function of Ω, and $R(z,\zeta)$ the Poisson kernel of Ω. Therefore we have for V the stronger properties (than the Nevanlinna class):

$$\lim_{\varepsilon\to 0} V_{|\partial\Omega_\varepsilon} = 2 \operatorname{Re} u , \quad \text{in } L^1(\partial\Omega).$$

REFERENCES.

[1] L. Carleson: The Corona theorem . (Proceedings of the 15th Scandinavian Congress, Oslo 1968), Lecture Notes in Mathematics, 118, Springer, Berlin-Heidelberg-New-York, 1970, 121 - 132.

[2] G.B. Folland and J.J. Kohn: The Neumann Problem for the Cauchy-Riemann complex. Annals of Mathematic Studies, 75, Princeton University Press, 1972.

[3] G.B. Folland and E.M. Stein: Estimates for the $\bar{\partial}_b$-complex and Analysis on the Heisenberg group. Communication on Pure and Applied Mathematics, 27, 1974, 429 - 522.

[4] L. Gruman: The zeros of holomorphic functions in strictly pseudoconvex domains (to appear in Trans. Amer. Math Soc.)

[5] G.M. Henkin: Integral representations of functions in strictly pseudoconvex domains and application to the $\bar{\partial}$-problem. (In Russian) Math. Sb. 82, 124, 1970, 300 - 308. English translation in Math. U.S.S.R - Sb. 11, 1970, 273 - 88.

[6] G.M. Henkin: Preprint, to appear in Doklady Ak. Nauk U.S.S.R., 1975.

[7] L. Hörmander: Generators for some rings of analytic functions. Bull. Amer. Math. Soc., 73, 1967, 943.

[8] L. Hörmander: L^p estimates for plurisubharmonic functions. Math. Scand. 20, 1967, 65 - 78.

[9] N. Kerzman: Hölder and L^p estimates for the solution of $\bar{\partial}u = f$ in strongly pseudoconvex domains. Comm. Pure and Appl. Math. 24, 1971.

[16] G. Laville: Résolution du $\partial\bar{\partial}$ avec croissance dans les ouverts pseudoconvexes étoilés de $\mathbb{C}^n$. C.R. Acad. Sc. Paris, 274, 1972, A - 554 - 556.

[17] G. Laville: Diviseurs et classe de Nevanlinna. Thèse de 3è cycle, Université de Paris VI, Juin 1975.

[18] P. Lelong: Fonctionelles analytiques et fonctions entières (n variables). Montrèal, les Presses de l'Université de Montrèal, 1968 (Séminaire de Mathematiques Superieures, été 1967, no 28).

[19] P. Lelong: Fonctions plurisousharmoniques et formes differentielles positives. Paris-London-New-York, Gordon and Breah, Dunod, 1968.

[20] I. Lieb: Das Ramirezsche Integral und die Lösung der Gleichung $\bar{\partial}f = \alpha$ im Bereich der beschränkten Formen. William Marsh Rice University Houston, Texas, 56, no 2, Spring 1970.

[21] P. Malliavin: Fonctions de Green d'un ouvert strictement pseudoconvexe et classe de Nevanlinna. C.R. Acad. Sc. Paris, 278, 1974, A-141 - 144.

[22] N. Øvrelid: Integral representation formulas for differential forms and solutions of the $\bar{\partial}$-equation. Colloque international du C.N.R.S. no 208: Fonctions analytiques de plusieures variables et analyse complexe. Agora Mathematica, Paris, Gauthier-Villars, 1974.

[23] H. Skoda: Valeurs au bord pour les solutions de l'opérateur d" dans les ouverts strictement pseudoconvexes. C.R. Acad. Sc. Paris, 280, 1975, A-633 - 636.

[24] H. Skoda: Zéros des fonctions de la classe de Nevanlinna dans les ouverts strictment pseudoconvexes. C.R. Acad. Sc. Paris, 280, (23 Juin 1975), A-1677 - 1680.

[25] H. Skoda: Valeurs au bord pour les solutions de l'opérateur d" et caracterisation des zeros des fonctions de la classe de Nevanlinna. Preprint, Centre Universitaire de Toulon et du Var.

ON A CLASS OF BANACH ALGEBRAS

by N. Th. Varopoulos

I would like to describe here certain algebras of analytic functions that have obvious connections with operator theory on a Hilbert space. The results that I have are very partial but some of the problems that arise are interesting.

The algebras in question were first pointed out to me in their commutative version by B. Cole (private communication). To give a formal definition let $n \geq 1$ be a positive integer and let $P_n = C[z_1, z_2, \ldots z_n]$ be the algebra of complex polynomials in n variables. We shall define a norm $\| \ \|$ on P_n by setting :

(1) $$\|p\| = \sup \|p(T_1, T_2, \ldots T_n\|_{op} \qquad p \in P_n$$

where we denote by $\| \ \|_{op}$ the Hilbert space operator norm of the operator $p(T_1, T_2, \ldots, T_n)$ and where the $\sup$ in (1) is taken over all collections of operators $T_1, T_2, \ldots, T_n \in \mathcal{L}(H)$ acting on some fixed separable (infinite dimensional)

Hilbert space H and satisfying :

(2) $\qquad : \qquad T_iT_j = T_jT_i \qquad i,j = 1, 2, \ldots, n.$

It is an easy matter to verify that the norm defined by (1) and (2) is submultiplicative and that we can complete P_n with the above norm to obtain a Banach algebra which we sahll denote by A_n.

Analogously an algebra A_∞ can be obtained by assigning on the ring $C[z_1, z_2, \ldots]$ (the ring of complex polynomials with countably many variables) a norm as in (1) and (2) and then by completing this ring. B. Cole has proved the following

THEOREM 1 (B. Cole).- The algebras A_n $(n = 1, 2, \ldots, \infty)$ are isometrically Hilbert space operator algebras.

We mean by that, of course, that for all n there exists $\mathfrak{A}_n \subset (H)$ a closed subalgebra of operators on the Hilbert space H such that $\mathfrak{A}_n = A_n$ isometrically. The proof will be omitted here . (For a non commutative version cf. [1]).

One fact that is easy to show also is that every commutative subalgebra $\mathfrak{A} \subset (H)$ of operators on H can be identified isometrically with a quotient subalgebra of A_∞ (over a closed ideal).

Now the main "result" that I have here is the following

THEOREM 2.- A_n is a semi simple algebra for all n.

This is an easy and formal result, we shall outline its proof at the end. If we combine Theorem 2 with our remark just above we obtain at once :

THEOREM 3.- There exists a semi simple commutative closed subalgebra of (H) that is not a Q-algebra.

(Cf. [2] for definitions and background on Q-algebras). Indeed A_∞ is not a Q-algebra since we know (Cf. [3]) that there exists at least one Hilbert space operator algebra that is not a Q-algebra.

What Theorem 2 says really is that we can identify A_n with an algebra of analytic functions on $D^n = \{(z_1, z_2, \ldots, z_n) \in C^n \; ; \; |z_i| \leq 1, \; i = 1, 2, \ldots, n\}$ $(n \geq 1)$ and that $\|f\| \geq \|f\|_\infty$ for all $f \in A_n$ (we denote by $\| \;\|$ the norm in A_n and by $\| \;\|_\infty$ the uniform norm in D^n).

The algebra A_n is therefore none other than the algebra of analytic functions that "operate" on n-families of commuting contractions on a Hilbert space.

It is a consequence of classical results of J. Von Neumann and T. Ando (cf. Ch. I of [4]) that the algebra A_n is isometrically isomorphic with $A(D^n)$ (the algebra of analytic functions continuous up to the boundary) for $n = 1, 2$. It is a consequence of [5] [6] that A_3 is not isometrically isomorphic with $A(D^3)$.

Whether $A(D^n) = A_n$ up to norm equivalence for $n \geq 3$ I do not know. More precisely we have $A(D^n) = A_n$ up to norm equivalence for some $n \geq 1$ if and only if there exists K_n some constant depending on n anly such that :

$$\|p(T_1, T_2, \ldots, T_n)\|_p \leq K_n \|p\|_\infty \qquad p \in P_n$$

and all family of operators $T_1, T_2, \ldots T_n$ satisfying (2) ; this is an open problem (one can see by analysing the proof in [5] that K_n if it exists has to satisfy $K_n \geq C\, n^\alpha$ for some $\alpha > 0$).

Just to indicate how weird the norm $\| \ \|$ is I state bellow (without proof) the following

THEOREM 4. Let $f(z_1, z_2, \ldots z_n) \in A_n$ then
$\|z_1^{\nu_1} z_2^{\nu_2} \ldots z_n^{\nu_n} f\| \to \|f\|_\infty$ as $\nu_1, \nu_2, \ldots \nu_n \to \infty$ (indep.)

There is a non commutative analog of the algebras A_n and A_∞ which is obtained by norming and completing appropriately the free algebra on n generators (rather than the free commutative algebra P_n). These algebras have already been considered in [1]. Theorem 2 is also valid for these algebras, but the proof being more intricate will be omitted.

We shall give finally :

Outline of the proof of Theorem 1 :

To simplify notations let us suppose that n is fixed and is finite, let us define then

$$\tau_p^\theta : A_n \to A_n \qquad 1 \leq p \leq n \qquad \theta \in R \pmod{2\pi}$$

a linear mapping which is defined by :

$$\tau_p^\theta f(z_1, z_2, \ldots z_n) = f(z_1, z_2, \ldots, z_p e^{i\theta}, \ldots z_n) \tag{3}$$

for all $f \in P_n$. It is easy to verify that τ_n^θ as defined on P_n in (3) is norm decreasing for the norm $\| \ \|$ of A_n and therefore extends to a linear mapping on A_n. It is also easy to verify that τ_p^θ is continuous in θ (for fixed p) in the sense that $\tau_p^\theta f \to f$ in A_n for all fixed p and $f \in A_n$ as $\theta \to 0$.

Let now $f \in A_n$ be fixed and let us define for every $m_1, m_2, \ldots m_n \geq 0$ the Fourier coefficient

$$\hat{f}(m_1,\dots m_n) = \int \tau_1^{\theta_1} \tau_2^{\theta_2} \dots \tau_n^{\theta_n} f\, e^{-i(m_1\theta_1+\dots+m_n\theta_n)} d\theta_1 \dots d\theta_n \in A_n$$

(for the normalized measure). Standard methods (cf. [7]) show that if $\hat{f}(m_1,\dots m_n) = 0$ for every non negative multiindex $(m_1, \dots m_n)$ then $f = 0$ in A_n. It is also obvious that if f belongs to the radical of A_n so does every coefficient $\hat{f}(m_1,\dots m_n)$ of f.

Now for every $f \in A_n$ and every multiindex $(m_1 \dots m_n)$ we have :

(4) $$\hat{f}(m_1,\dots, m_n) = a_{m_1\dots m_n} z_1^{m_1} \dots z_n^{m_n}$$

for some scalar $a_{m_1\dots m_n} \in \mathbb{C}$. This is obvious if $f \in P_n$ and follows for every $f \in A_n$ by continuity. It is an easy matter to verify now that an element of A_n as in (4) belongs to radical of A_n if and only if $a_{m_1,\dots,m_n} = 0$. This completes the proof of the Theorem.

It is the very last step that has to be modified somewhat in the non commutative case because then (4) does not hold.

REFERENCES.

[1] N.Th. Varopoulos: A Theorem on operator algebras. To appear (Math. Scand.)

[2] A.M. Davie: Quotient algebras of uniform algebras. J. London Math. Soc. 7 (1) 1973, p.p. 31 - 40.

[3] N.Th. Varopoulos: Sur une inegalite de Von Neumann. C.R. Acad. Sci. Paris 277 (A), p.p. 19 - 22.

[4] B.Sz. Nagy & C. Foias: Harmonic analysis of operators on Hilbert spaces. North Holland 1970.

[5] N.Th. Varopoulos: On an Inequality of Von Neumann and an application of the metric theory of tensor products to operator theory. J. of Func. Analysis 16 (1) 1974, p.p. 83 - 100.

[6] A.M. Davie & M.J. Crabb (preprint).

[7] Y. Katznelson: An Introduction to Harmonic Analysis. J. Wiley (1968)

Pseudodifferential Operators and the $\bar{\partial}$-Equation

by Nils Øvrelid *)

0. Introduction: The present note reproduces with small changes talks given at Agder Distriktshøgskole and the Nordic Summer School in Mathematics at Grebbestad. It was intended as a preliminary report on the regularity of Kohn's solution. We recall that the Kohn solution $K(f)$ to $\bar{\partial}u = f$ is the unique solution orthogonal to $\ker \bar{\partial}$. In the strictly pseudoconvex case, it is known that $K(f)$ has optimal regularity in terms of Sobolev (H_s) spaces, and for some time this has been suspected for other norms as well. Thus, one might hope for a gain of $\frac{1}{2}$ derivative in all directions and 1 along the complex tangent spaces. We show that this is always true up to ε for every $\varepsilon > 0$, and in some cases exactly. However, this question of sharp estimates is not treated systematically. On the other hand, recent counterexamples [11], [13] show that the sharp estimates are the best one can hope for.

Note on related work: At the AMS Summer Institute at Williamstown, I recently learnt about the closely related work of Greiner and Stein [5], [6]. They use a somewhat different method, based on approximating the boundary by the Heisenberg group, and for suitably adapted metrics their results seem to imply those of the present paper, and even to settle most questions about sharp estimates. It is also likely that the restrictions on the metric may be removed by using recent work of Rotschild and Stein [15]. Still, it is hoped that the present approach might be of interest. It should also be mentioned that the recent construction by Boutet de Monvel, Grigis and Helffer of parametrices in S_Σ [3] is methodically very close to that of section 4.

Acknowledgement: I am indepted to Directors and Fellows of l'Institut Mittag-Leffler, in particular Lars Hörmander, for valuable discussions and advice.

1. Preliminaries. We use the basic notation of [7], and let $\lrcorner$ denote the interior product or contraction: $V \times \Lambda^q V \to \Lambda^{q-1}V$ determined by a scalar product on V. Consider the following setting:

I. D is a relatively compact domain with smooth boundary ∂D in a complex manifold M of dimension $n > 1$. M is given a smooth Hermi-

*) This paper was prepared at the Mittag-Leffler Institute while the author was on leave from Oslo University, with financial support from the Mittag-Leffler Foundation.

tean metric and the corresponding volume measure.

II. Let q be an integer in $[1,n-1]$, and consider the $\bar{\partial}$-equation for $(0,q)$-forms in D. Suppose $r \in C^{\infty}(M)$ defines D in the sense that $D = \{r<0\}$ and $dr \neq 0$ on ∂D. The complex tangent bundle to ∂D, $T^{c}\partial D = \ker(\partial r : T\partial D \to \mathbb{C})$.

Basic assumption: The Leviform $Lr|_{T^{c}\partial D}$ is everywhere nondegenerate with index (= #neg.eigenvalues) $\neq q$.

In section 4, remark, we see that this assumption may be weakened to Condition Z(q) ([4] p.57):

$Lr|_{T^{c}_{z}\partial D}$ has at least $q+1$ negative or at least $n-q$ positive eigenvalues at every $z \in \partial D$.

Under these assumptions, we shall study the regularity of Kohn's solution for $(0,q)$-forms.

2. The $\bar{\partial}$-Neumann problem: Let $f \in L^{2}_{(0,q)}(D)$ satisfy $\bar{\partial} f = 0$. If $\bar{\partial} u = f$ is solvable in L^{2}, the Kohn solution or canonical solution $K(f)$ is the unique solution orthogonal to $\ker \bar{\partial}$. Under assumptions I and $Z(q)$; $K(f)$ may be described by the $\bar{\partial}$-Neumann problem [4]. The $\bar{\partial}$-Neumann boundary conditions are

(2.1) $\quad \bar{\partial} r \lrcorner u = 0$; $\bar{\partial} r \lrcorner \bar{\partial} u = 0$ on ∂D.

As usual,

$\square = \theta\bar{\partial} + \bar{\partial}\theta$ is the complex Laplacian on $(0,q)$-forms.

We have:

(a) $\quad \|u\|_{(s+1)} \leq C_{s}(\|\square u\|_{(s)} + \|u\|_{(0)})$

when $s \geq 0$ and $u \in C^{\infty}_{(0,q)}(\bar{D})$ and satisfies (2.1).

(b) Let F be the unique closed, selfadjoint extension of $\square$ corresponding to boundary condition (2.1), and N a left inverse to F, modulo $\bar{\partial}$-harmonic forms. Then $K(f) = \theta N f$ whenever $\bar{\partial} u = f$ is solvable.

Any parametrix to the $\bar{\partial}$-Neumann problem will only differ from N by an operator that is smoothing up to the boundary, and thus allows us to study the regularity of Kf.

3. Reduction to the boundary. We shall use the standard method of reducing a boundary value problem to a pseudodifferential equation on the boundary, as developed in [8]. Pseudodifferential operators are called

classical if they are in the algebra of Kohn-Nirenberg [12] (or [8]), otherwise they need only be of type $(\frac{1}{2},\frac{1}{2})$ in the sense of [9]. Let $\equiv$ denote equality modulo terms smooth up to the boundary in D. Pick a parametrix T to $\Box$, and let Q_0, Q_1 be the double layer and single layer operators entering in Green's formula. They map $\Gamma(\partial D, \Lambda^{0,q})$ to $C^\infty_{(o,q)}(D)$, $\Box \cdot Q_1 \equiv 0$, and the operators $Q_{kl} : v \to \lim_{D \ni x \to \partial D} D_n^k(Q_l v(x))$, $l = 0,1$; are classical pseudos on ∂D, with symbols computable from [8].

Write the boundary problem as

$$(3.1) \qquad \Box u = f, \quad B(\gamma_0 u, \gamma_1 u) = 0,$$

where $\gamma_0 u$, $\gamma_1 u$ are the boundary values from the interior of u and its normal derivative, and B is a first order operator on ∂D. Pick a solution of $\Box v = f$ with optimal boundary behavior, and solve

$$(3.2) \qquad B(w_0, w_1) = -B(\gamma_0 v, \gamma_1 v)$$

$$(3.3) \qquad w_1 = Q_{10} w_0 + Q_{11} w_1$$

up to smooth remainders. Then $w_0 \equiv Q_{00} w_0 + Q_{01} w_1$ follows automatically, and $u = v + Q_0 w_0 + Q_1 w_1$ solves (3.1) up to smooth remainders. Since the principal symbol $\sigma(Q_{11}) = \frac{1}{2}\cdot \mathrm{id}$, $I - Q_{11}$ has a parametrix Q', and we replace (3.3) by

$$(3.4) \qquad w_1 = Q' Q_{10} w_0 .$$

As usual, we need only find a pseudolocal parametrix to (3.2) - (3.4) near each point. Let $\bar\omega^1, \ldots, \bar\omega^n$ be a local, orthonormal frame of (0,1)-forms, and $\bar Z_j = 1/\sqrt{2}[X_{2j-1} + iX_{2j}]$, $j = 1, \ldots, n$ the reciprocal frame of antiholomorphic vector fields, with $\bar\omega^n$ proportional to $\bar\partial r$ and X_{2n} = the unit normal field to ∂D. Writing

$$u = \sum_{\substack{|j|=q-1 \\ n \notin j}}' u'_j \bar\omega^n \wedge \bar\omega^j + \sum_{n \notin I}' u_I \bar\omega^I = \bar\omega^n \wedge u' + u''$$

we get

$$\bar\omega^n \lrcorner\, u = u'$$

$$\bar\omega^n \lrcorner\, \bar\partial u = \sum_{n \notin I}' (\bar Z_n u''_I)\bar\omega^I + B_0 u'' + B_1 u'$$

where B_0, B_1 are operators of order 0 and 1, and the boundary equation $B(w_0, w_1) = (g,h)$ takes the form

$$(3.5) \qquad w'_0 = g$$

$$(3.6) \qquad \sum_{n \notin I}' 1/\sqrt{2}[X_{2n-1} w_{0,I} + i w_{1,I}]\bar\omega^I + B_0 w''_0 + B_1 w'_0 = h .$$

Substituting (3.3) and (3.5) in (3.6), we get

$$(3.7)\qquad A\cdot(w_{0,I})_{n\notin I} = C(f,g)\,, \quad \text{where}$$

$$A = A_1\cdot I_E + A_0\,,$$

with A_1 a scalar, classical pseudodifferential operator of order 1, and A_0 a classical pseudodifferential $E\times E$-matrix of order 0; $E = \binom{n-1}{q}$.

It is clear that a parametrix for A gives a parametrix for the system (3.3) - (3.5) - (3.6).

<u>4. The parametrix construction.</u> Let J be multiplication by $i = \sqrt{-1}$ in TM. For each $x\in\partial D$, the principal symbol of A, $\sigma(A)(x,\xi) = (|\xi| - \xi_{2n-1})\cdot I$ in coordinates suitably adapted to x, and $\sigma(A)$ vanishes to exactly the second order on the characteristic manifold $\Sigma = \{(z, t\cdot dr(z)\circ J) : z\in\partial D \text{ and } t<0\}$. The existence of a parametrix in Boutet de Monvel's class $S_\Sigma^{-1,-2}$ follows easily from the general theory of [2], but this does not seem precise enough, and we shall proceed differently. Let $\theta_j = \sigma(X_j)$, $j = 1,\ldots,2n-2$, in the local coordinates, and let τ be a supplementary ξ-variable. A function $\varphi(s,t)$, $s\in\mathbb{R}^{2n-2}$ and $t\in\mathbb{R}$, is said to be quasihomogeneous of type (1,2) and weight m if $\varphi(\lambda^{\frac12}s,\lambda t) = \lambda^m\varphi(s,t)$ for all $\lambda>0$.

Quasihomogeneous multiplier operators are (roughly) equally good on all spaces, and the main idea in the parametrix construction is to enlarge the classical pseudodifferential algebra by symbols quasihomogeneous of type (1,2) in (θ,τ).

When a and b are symbols,

$$a\circ b \sim \sum_\alpha \frac{(-i)^{|\alpha|}}{\alpha!}(\partial/\partial\xi)^\alpha a\cdot(\partial/\partial x)^\alpha b$$

denotes the symbol of (operator of a) $\circ$ (operator of b).

Counting θ with weight $\frac12$ and τ with weight 1, we write

$$\text{symbol}(A) = a = \Big(\sum_{j,k=1}^{2n-2} a_{jk}(x)/\tau\;\theta_j\,\theta_k\Big)I + a_0(x) + \text{terms of neg.weight, on a}$$

conical neighbourhood U of Σ, and observe that $a\circ b(x,\theta/\tau^{\frac12}) = Q_x(s,D_s)b(x,s)|_{s=\theta/\tau^{\frac12}} + \text{neg.weight}$, where Q_x is a second order differential operator depending smoothly on x. In suitable local frame and coordinates, we get

$$Q_x(s,D_s) = \sum_{j=1}^{n-1} [D^2_{2j-1} + D^2_{2j} + 2\epsilon_j (s_{2j}D_{2j-1} - s_{2j-1}D_{2j}) + s^2_{2j-1} + s^2_{2j}]\cdot I + a_o(x)\,; \qquad \text{with } \epsilon_j = \pm 1 .$$

With $B_N = \{f : x^\beta D^\alpha f \in L^2 \text{ when } |\alpha|+|\beta| \leq N\}$, such operators are Fredholm: $B_{N+2} \to B_N$; $N \geq 0$. (E.g. [1], section 7.)
Composing a by $b = \tau\cdot(\sum_{j,k} a_{jk}(x)\theta_j\theta_k\cdot I + b(x)\tau)^{-1}$ and iterating, we may find rationel semihomogeneous symbols p'_N, for each $N \geq 0$, such that $a\circ p'_N = I - r_N(x,\theta/\tau^{\frac{1}{2}}) + r'_N$, with weight $r'_N \leq -\frac{1}{2}$ and $r_N(x,\cdot) \in B_N^{E\times E}$. Suppose $Q_x : B^E_{N+2} \to B^E_N$ is bijective for each x and N. Then we solve $Q_x p''_N(x,s) = r_N(x,s)$, and see that $p_N(x,\xi) = p'_N(x,\theta,\tau) + p''_N(x,\theta/\tau^{\frac{1}{2}})$ is a right inverse to a in U modulo symbols of negative weight, and it is of any prescribed order of smoothness when N is large enough. The case of a left parametrix is handled similarly.
In [10], Hörmander shows that hypoellipticity with loss of one derivative implies bijectivity from B_{N+2} to B_N of certain test operators obtained by localizing. Q_x arises in this way, and the "basic estimate" (2.a) implies a similar estimate for A. At least in the Kähler case, it is possible to give a more selfcontained treatment by computing the spectrum and its forbidden values for $a_o(x)$, following the analysis in [2] and [10].

Remark: By using recent results of Boutet de Monvel, Grigis and Helffer [3] it is possible to replace the condition of nondegenerate Leviform by Condition Z(q), and still obtain $p_N(x,\xi)$ with the properties above.

5. Estimates. Under the assumptions of Section 1, we have found right and left inverses under O to a in a conical domain U, modulo operators of negative weight. We microlocalize the problem, i.e. introduce a smooth cut-off function $\chi(\xi)$, homogeneous of order zero, with $\chi = 1$ near Σ and $\chi = 0$ outside U. The corresponding multiplier M_χ preserve L^p-spaces; $1 < p < \infty$, as well as Lipschitz spaces (e.g. [14]). Outside U we have a classical parametrix $\tilde{P} : \tilde{P}\circ A\circ(I-M_\chi) = (I-R_\chi)\circ(I-M_\chi)$ etc.

Then we must study $p_N\circ\chi$ and remainder $\circ\,\chi$ and need to know the regularity properties of operators with quasihomogeneous symbols

$b(x,\theta,\tau)$ of weight $m = 0$ or $-\frac{1}{2}$. Inverse Fourier transform gives integral operators $k(x,z)$, highly differentiable when $z \neq 0$, and the dependence on x is so mild that they have the same regularity as convolution operators quasihomogeneous in (z', z_{2n-1}) of weight $n - m$. With the notation of [16] Ch. V for Lipschitz spaces, we get the operators $OP(b)$ continuous from Λ_α to $\Lambda_{\alpha+\frac{1}{2}}$ when $m = -\frac{1}{2}$, $0 \leq \alpha < \frac{1}{2}$, while $OP(b)$ is at least continuous Λ_α to $\Lambda_{\alpha'}$ when $0 \leq \alpha' < \alpha$, $m = 0$. (According to N. Riviere (personal communication), the continuity $\Lambda_\alpha \to \Lambda_\alpha$ may fail in this case.)

Now we use the symbol calculus for pseudodifferential operators, writing the symbols occurring as sums $a_o + \sum_{j=1}^{N} a_j b_j a_j' + c$, where a_o, a_j, a_j' are classical symbols, b_j is quasihomogeneous of order 0 or $-\frac{1}{2}$, and c has low order. Then we count the mixed order of the mixed terms as order a_j + order a_j' + weight b_j. Then operators of negative order must always map $L^\infty \to L^\infty$ continuously. The fact that we have introduced symbols of finite degree of smoothness M does not matter, as long as M is large compared with the number of compositions and the order of regularity of function spaces we consider. The existence of a (left or right) parametrix P, modulo an operator regularizing in L^p_s and Λ_α, follows. [Operator of neg.order] $\circ$ P map L^∞ to L^∞, and from consideration of P^* also L^1 to L^1 continuously, and we may interpolate and conjugate by Riez potentials.

We substitute in the formula for (the approximate) Neumann operator, and apply θ. To the worst terms, we only apply vector fields along the complex tangent spaces, and the mixed order of the boundary operator $P \circ [\text{order} - 1]$ goes up by only $\frac{1}{2}$. If we apply to $K(f)$ a vector field X with $X_z \in T_z^c \partial D$ when $z \in \partial D$, only another increase by $\frac{1}{2}$ results. This leads to <u>the coarse estimates</u>.

<u>Theorem 5.1</u> Make the preceeding assumptions. For every $\varepsilon > 0$ we have continuity of

(1) $K : \Lambda_{\alpha(o,q)}(D) \to \Lambda_{\alpha+\frac{1}{2}-\varepsilon(o,q-1)}(D)$ and

$X \circ K : \Lambda_{\alpha(0,q)}(D) \to \Lambda_{\alpha-\varepsilon,(o,q-1)}(D)$, when $\alpha > 0$.

(2) $K : L^p_{s(o,q)}(D) \to L^p_{s+\frac{1}{2}-\varepsilon(o,q-1)}(D)$; $1 < p \leq \infty$, $s \geq 0$ or $s > 0, p = 1$,

and $X \circ K : L^p_{s(o,q)}(D) \to L^p_{s-\varepsilon(o,q-1)}(D)$, when $1 \leq p \leq \infty$ and $s > 0$.

In the formula for K, one sees that in the top order terms of the symbol of the boundary operator, there is a semihomogeneous factor of order $-\frac{1}{2}$. This leads to the sharp estimates

<u>Theorem 5.2</u> K is continuous $C^k_{(o,q)}(\bar{D}) \to \Lambda_{k+\frac{1}{2}(o,q)}(D)$, and from $\Lambda_{\alpha(o,q)}(D)$ to $\Lambda_{\alpha+\frac{1}{2}(o,q-1)}(D)$, at least when $\alpha, \alpha+\frac{1}{2} \notin \mathbb{Z}$.

The corresponding statement about the boundary symbol of $X \circ K$ is not true.

References.

1. R. Beals: A general calculus of pseudodifferential operators. Duke Math. Journal 42 (1975), 1-42.

2. L. Boutet de Monvel: Hypoelliptic operators with double characteristics and related pseudo-differential operators. Comm. Pure Appl. Math. 27 (1974), 585-639.

3. L. Boutet de Monvel, A. Grigis and B. Helffer: Parametrices to hypoelliptic operators with double characteristics. Talks at Grebbestad 1975.

4. G.B. Folland and J.J. Kohn: The Neumann problem for the Cauchy-Riemann complex. Princeton University Press 1972.

5. P.C. Greiner and E.M. Stein: A parametrix for the $\bar{\partial}$-Neumann problem. To appear in Proceedings of Rencontre sur plusieurs variables complexes et le problème de Neumann, Montreal 1974. Presses Universitaires de Montreal 1975.

6. P.C. Greiner and E.M. Stein: Regularity for the $\bar{\partial}$-Neumann problem in strongly pseudoconvex domains. Report to 1975 A.M.S Summer Institute on Several Complex Variables.

7. L. Hörmander: L^2 estimates and existence theorems for the $\bar{\partial}$-operator. Acta Math. 113 (1965), 89-152.

8. ——— : Pseudo-differential operators and non-elliptic boundary problems. Annals of Math. 83 (1966), 129-209.

9. ——— : Pseudo-differential operators and hypoelliptic equations, in Proc. Symp. Pure Math. vol.10. Am. Math. Soc. 1966.

10. ——— : A class of hypoelliptic pseudo-differential operators. To appear.

11. N. Kerzman: Hölder and L^p- estimates for solutions of $\bar{\partial}u = f$ in strongly pseudoconvex domains. Comm. Pure Appl. Math. 24 (1971), 301-380.

12. J.J. Kohn and L. Nirenberg: On the algebra of pseudodifferential operators. Comm. Pure Appl. Math. 18 (1965), 269-305.

13. S. Krantz: Optimal Lipschitz and L^p estimates for the equation $\bar{\partial}u = f$ on strongly pseudo-convex domains. To appear.

14. N. Riviere: Class of smoothness, the Fourier method. Unpublished notes.

15. L.P. Rothschild and E.M. Stein: Hypoelliptic differential operators and nilpotent groups. To appear.

16. E.M. Stein: Singular integrals and differentiability properties of functions. Princeton University Press 1970.

PROBLEMS.

1.

Let D_1, D_2 be Jordan domains such that

$$\iint_{D_1} z^n \, dxdy = \iint_{D_2} z^n \, dxdy \qquad n = 0, 1, 2, \dots .$$

Must we have $D_1 = D_2$?

Remark: Answer is yes if $\overline{D}_1 \cap \overline{D}_2$ is empty, or consists of just one point. I suspect the answer is no in general. The problem can also be formulated thus: The Cauchy transform of a bounded domain D is defined as

$$S_D(\zeta) = \iint_D \frac{dxdy}{z-\zeta} \qquad (z = x + iy \; ; \; \zeta \in \mathbb{C} \setminus \overline{D}).$$

Can two distinct Jordan domains have Cauchy transforms which are equal for large $|\zeta|$?

H. Shapiro

2.

Construct a pseudomeasure μ on the circle Π such that $\mu \neq 0$ and

(i) $\operatorname{supp} \mu$ is of Lebesgue measure zero

(ii) $\sum_{n=0}^{\infty} |\hat{\mu}(n)|^2 < \infty$.

Remarks: A measure μ satisfying (i) and (ii) must be the zero measure ; this is an immediate consequence of the F. and M. Riesz theorem.

In other terms, the problem can be stated thus: Construct two non-constant analytic functions f, g such that

(a) f is analytic in $|z| < 1$ and in the Hardy class H^2

(b) g is analytic in $1 < |z| \leq \infty$ and has bounded Taylor coefficients (referring to the expansion in powers of z^{-1})

(c) f and g are analytic continuations of one another across each point z_0, $|z_0| = 1$ except for a closed subset of the circle of Lebesgue measure zero.

Presumably examples can even be constructed where g has Taylor coefficients that are $O(\frac{1}{\sqrt{n}})$, etc.

H. Shapiro

3.

It is known that a trigonometric series of class L^2, with Hadamard gaps, that vanishes on a set of positive measure, vanishes identically. That is, $\sum_{-\infty}^{\infty} |c_n|^2 < \infty$, $f(\theta) \sim \sum_{-\infty}^{\infty} c_n e^{in\theta}$, $c_n = 0$ for all $n \neq \pm n_k$ where $1 \leq n_1 < n_2 < \dots$, and $\inf \frac{n_{k+1}}{n_k} > 1$ and $f(\theta) = 0$ for θ in a set of positive measure imply $f = 0$.

Question: Is the same conclusion valid if f has Hadamard gaps on one side only, that is, assuming only $c_n = 0$ whenever $n > 0$, $n \neq n_k$? In complex-variable terms the problem is stated thus: Can we find nonconstant analytic functions f, g which are, respectively, in the Hardy classes H^2 of the interior and exterior of the unit disc, moreover f is Hadamard-lacunary, and which have equal boundary values on a subset of the circle having positive measure?

H. Shapiro

4.

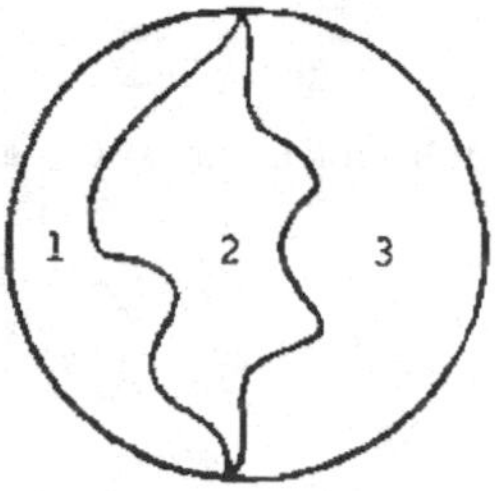

"Three-countries problem": The unit disc is devided into 3 countries, as depicted, by means of two smooth Jordan arcs running from the north pole to the south pole.

Must there exist a function harmonic in the disc, which is negative on country 1, and positive on country 3 ?

Conjecture: In general no.
(This question arose in an attack on # 1).

H. Shapiro

PS. Arne Stray pointed out answer is yes by a simple application of Arakelian's approximation theorem. Still open is question: Can an integrable harmonic function be found with the above properties ; and for this I conjecture no in general.

5.

This is not a "serious" problem - it is offered for your amusement. It can be proved in 3 lines, and I promise a beer to anyone who solves it:

<u>Prove</u> f meromorphic in $\mathbb{C}$,
$f(z) = \bar{z}$ for all z on some arc Γ

$\Rightarrow$ f is a linear fractional function, and Γ is the arc of a "circle"*.

*"circle" means, circle on the Riemann sphere, i.e. a circle or straight line in $\mathbb{C}$.

H. Shapiro

6.

Let B denote the "Bergman space" of the unit disc $\mathbb{D}$, i.e. $\{f$ analytic in D : $\|f\|^2 = \frac{1}{\pi} \iint_D |f(z)|^2 \, dxdy < \infty\}$.

$f \in B$ is said to be <u>weakly invertible</u> if $\exists$ polynomials p_n such that $\|p_n f - 1\| \to 0$. (This is the B-analogue of an outer function).

General problem: Find an "effective" NASC for f to be weakly invertible.

<u>Specifically</u>: Is it true that $f \in B$ and

(i) $|f(re^{i\theta})| \geq C(1-r)^N$

for some positive C, $r \Rightarrow f$ is w.i. in B ?

<u>Remark</u>: It is known that (i) together with $\iint_D |f(z)|^{2+\varepsilon} dxdy < \infty$ some $\varepsilon > 0$ is <u>sufficient</u> for f to be w.i.

H. Shapiro

7.

Let D be a bounded simply-connected domain. Do finite linear combinations of the functions $z \to (z-\zeta)^{1/2}$, $\zeta \in \partial D$ span the space $L_a^1(D)$ (analytic functions in D, integrable with respect to planar measure) ?

I can prove yes if D is star-shaped, suspect answer is no in general.

Do finite linear combinations of functions of the form

$$z \to (z-\zeta)^p , \quad \zeta \in \partial D$$

where p ranges over all positive real values, span $L^1_a(D)$?

I can prove yes when ∂D satisfies some regularity conditions - maybe it is always true ? (The motivation was to exhibit explicit families of bounded functions which are dense in $L^1_a(D)$).

H. Shapiro

8.

B : Bergman space on $\mathbb{D}$, as in problem # 6.
Conjecture: Let $E \subset \mathbb{D}$ be a zero-set for B, i.e. a set such that $\exists f \in B$, $f \not\equiv 0$, and $f(z) = 0$ for all $z \in E$ and for no other $z \in \mathbb{D}$. Then, there exists a function $g \in B$ vanishing on E such that the closure of its polynomial multiples contains all $f \in B$ which vanish on E.

H. Shapiro

9.

Let B be a closed subalgebra of L^∞ of the unit circle which contains H^∞. Let C_B be the C^*-algebra generated by the inner functions that are invertible in B. Is $B = H^\infty + C_B$? The equality is known to hold for all the algebras B that have been investigated in detail [1] [2] [3].

[1] A.M. Davie, T.W. Gamelin, J. Garnett: Distance estimates and pointwise bounded density, Trans. Amer. Math. Soc.
[2] S-Y. Chang: On the structure and characterization of some Douglas subalgebras, Amer. J. Math, to appear.
[3] D. Sarason: Functions of vanishing mean oscillation, Trans. Amer. Math. Soc., to appear.

D. Sarason

10.

Notations as in the preceding problem. Is it true that any unimodular function in C_B can be uniformly approximated by quotients of inner functions that are invertible in B?

The answer is known to be "yes" in the special cases that have been investigated in detail ; see [1], [2] above and [4] below. It is tempting to try to formulate and prove an abstract approximation theorem which would yield the result.

[4] R.G. Douglas and W. Rudin: Approximation by inner functions, Pacific J. Math. 31 (1969), 313 - 320.

D. Sarason

11.

What are the set of uniqueness for the algebra $QC = (H^{\infty}+C) \cap \overline{(H^{\infty}+C)}$? In particular, is a measurable subset of the unit circle which meets every arc in a set of positive measure a set of uniqueness?
The question was originally raised, for $H^{\infty} + C$ rather than for QC, by some nonempty subset of {L. Brown, L. Rubel, A. Shields}.
There is reportedly an example of Kahane of a dense open subset of the unit circle which is not a set of uniqueness for $H^{\infty} + C$.

D. Sarason

12.

If f is a continuous function on the unit circle whose conjugate function is continuous, is the best approximate of f in H^{∞} also continuous ?

D. Sarason

13.

Does every function in L^{∞} of the unit circle have a best approximate in $H^{\infty} + C$? (I haven't thought at all about this one, so I don't know how substantial it is).

D. Sarason

14.

Let U be an open set in $\mathbb{C}^n$ (not necessarily a domain of holomorphy), $z^0 \in U$.
Let $O(M) = \{f$ holomorphic on $U\}$.
Let $M \quad = \{f(z^0) = 0\}$.
The following theorem is true:

Theorem: The coord. fnals $z_1 - z_1^0, \ldots, z_n - z_n^0$ generate algebraically the ideal M.

Question: Prove this using the minimum of deep results from several $\mathbb{C}$-variables.
e.g. do not assume that the maximal ideal space of $O(U)$ is a Stein manifold.

G. Dales

15.

Let K be a polyn. convex set in $\mathbb{C}^n$, and $f \in C^\infty(K)$. Suppose $\bar\partial f = 0$ to infinite order on K.
Is f uniformly approximable by polynomials?

N. Sibony

16.

Let Ω open set in $\mathbb{R}^n$. $D = \sum_{i=1}^{\infty} a_i(x) \frac{\partial}{\partial x_i}$ a_i are in $C^\infty_{\mathbb{C}}(\Omega)$.
Let $g \in L^\infty(\Omega)$ and suppose g is invertible in $L^\infty(\Omega)$ and that $Dg \in L^2(\Omega)$.
Is it true that $D(\frac{1}{g}) = -\frac{1}{g^2} D(g)$.
This will solve the inner function pb.

N. Sibony

17.

Let F be a compact subset of $\mathbb{C}$ such that $\mathbb{C} \setminus F$ has exactly two components U_1 and U_2. Assume $(\overline{U}_i)^0 = U_i$.
Let $A(U_i) = \{f \in C(\overline{U}_i) : f|_{U_i} \text{ is analytic}\}$, $i = 1, 2$.
Let E have positive harmonic measure w.r.t. U_1. Show that

$$(*) \qquad \sup_{\substack{f \in A(U_2) \\ \|f\|_{U_2} \le 1}} \ \inf_{\substack{g \in A(U_1) \\ \|g\|_{U_1} \le 1}} \|f-g\|_E = 1$$

(Even a result like $(*)$ with "$=1$" replaced by "$\geq \delta$" for some pos. constant, would be interesting).

A. Stray

18.

Let B and D be as in problem 6. Let $F \subset D$ be relatively closed. Let B_{bF} denote the space $\{f \in B : f|_F \text{ is bounded}\}$. Give necessary and/or sufficient geometrical conditions on F such that

for each $f \in B_{bF}$ there are polynomials $\{p_n\}$ such that $\{p_n|_F\}_{n=1}^{\infty}$ is a bounded sequence and $\lim_{n\to\infty} \iint_D |f-p_n|^2 dxdy = 0$.
A related problem is to describe the B_{bF}-hull of F :
$\{z \in D : |f(z)| \le \|f\|_F \text{ for all } f \in B_{bF}\}$.

A. Stray

19.

Let φ be a positive increasing function on $[0,2\pi]$ (you may suppose φ continuous). Let

$$f(t) = e^{i\varphi(t)} \sim \sum_{-\infty}^{\infty} \hat{f}(n) e^{int}$$

and let Pf be its "analytic projection" $\sum_{0}^{\infty} \hat{f}(n) e^{int}$.

Conjecture: There exists an absolute constant $c > 0$ such that

(1) $\|Pf\|_2 \ge c\|f\|_2$ ($\|\cdot\|_2$ denotes norm in $L^2(0,2\pi)$).

Remark: (1) is known in the special case that φ is bounded, and if moreover $\varphi(2\pi) \le 2\pi$ even in the sharper form
$(|\hat{f}(0)|^2 + |\hat{f}(1)|^2)^{1/2} \ge c\|f\|_2$.

H. Shapiro

20.

Let $D \subset \mathbb{C}^{n+1}$, bounded, be defined by $\rho < 0$, $d\rho \neq 0$ on ∂D ($\rho \in C^{\infty}$). For $\zeta \in \partial D$ choose affine linear coordinates u_ζ, s.th. $u_\zeta(\zeta) = 0$ and $\frac{\partial \rho}{\partial u_\zeta^i}(\zeta) = 0$, $i = 1, \ldots, n$, $\frac{\partial \rho}{\partial u_\zeta^{n+1}}(\zeta) = 1$.

Set, for $m \ge 2$, fixed integer,

$$t_{\zeta,m} = \sum_{1 \le |\alpha| \le m} \frac{1}{\alpha!} (D^{\alpha}_{u_\zeta} \rho|_{u_\zeta=0}) u_\zeta^{\alpha} .$$

$W_{\zeta,m} = (u_\zeta^1, \ldots, u_\zeta^n, t_{\zeta,m})$ defines hol. coord. near ζ.
On the hyperplane $t_{\zeta,m} = 0$ we have

$$\rho(u_\zeta^1, \ldots, u_\zeta^n, 0) = L_\zeta^{(m)}(w) + O(|w|^{m+1})$$

where $w = (u_\zeta^1, \ldots, u_\zeta^n)$.

We say: D is pseudoconvex of strict type m at ζ if $\exists \gamma, c > 0$ s. th. (*) $L_\zeta^{(m)}(w) \ge \gamma |w|^m$ for $|w| \le c$.
D is uniformly pseudoconvex of strict type m if (*) holds for all $\zeta \in \partial D$.

<u>Question 1</u>: If D is pseudoconvex of strict type m at ζ, is there a local <u>holomorphic</u> peaking function at ζ? (There is a <u>continuous</u> peaking function, which is holomorphic on $D \cup \partial D - \{P\}$).

<u>Question 2</u>: If D is uniformly pseudoconvex of strict type m, is $\overline{D}$ intersection of domains of holomorphy? (This problem may be more accessible than the corresponding question for arbitrary domains of holomorphy with smooth boundary).

M. Range

21.

Let H be a separable Hilbert space and let $\mathfrak{A}$ be a reductive operator algebra on H (i.e. weakly closed, contains 1 and every invariant subspace of the algebra is reducing). Suppose T is a spectral operator in $\mathfrak{A}$. Decompose T canonically as follows: $T = S + Q$ where S is the scalar type operator and Q is a quasi-nilpotent operator commuting with S. Does $S \in \mathfrak{A}$?

<u>Remark</u>: A counter-example to this would give a counter-example to the reductive algebra problem. The conjecture may be simple to prove and/or known - I haven't tried to prove it.

N.P. Jewell

22.

Let T be a decomposable operator in the sense of [1] with spectral subspaces $\mathcal{E}(F)$ for F closed in $\mathbb{C}$. Suppose S is a linear operator which commutes with T.

<u>Conjecture</u>: $S\,\mathcal{E}(F) \subseteq \mathcal{E}(F)$.

[1] Theory of Generalized Spectral Operators - Colojoara and Foias.

<u>Remark</u>: The conjecture is yes if we assume S is continuous, and is not hard to prove.

N.P. Jewell

23.

Let E be a totally disconnected compact set in $\mathbb{C}^n$ of Hausdorff dimension < 2. Show that E is polynomially convex.

<u>Remark</u>: Vitushkin has constructed a totally disconnected compact set E in $\mathbb{C}^2$ of Hausdorff dimension 2 which has a full ball in its polynomially convexhull. He in effect raised the question above.

J. Wermer

24.

Can the space $A(D)$ be a quotient space (as a Banach space) of $L(H)$? $L(H)$ is the space of operators on a Hilbert space H. This is of course, related with Pełczyński's work.

N. Varopoulos

25.

Let B be a uniform algebra and let us suppose that $B \cong \mathbb{C}(X)$ (as Banach space topological $\cong$) as a linear space for some compact space X. What can be said about B? Are there any non trivial examples of such B ?

N. Varopoulos

26.

Let $a = (a_{ij})_{i,j=1}^{\infty}$ be infinite matrices normed by

$$\|a\| = \sup\{ \sum_{i,j=1}^{\infty} a_{ij}x_i y_j \mid \Sigma|x_i^2| \leq 1, \Sigma|y_i^2| \leq 1\}$$

and let us define

$$a \cdot b = (a_{ij}b_{ij})_{i,j=1}^{\infty}$$

the pointwise multiplicative on this space of matrices. We obtain then a Banach algebra and the above norm in submultiplicative. (This is non trivial but true).

The above algebra is an operator algebra, i.e. it is algebra $\cong$ to a closed subalgebra of $L(H)$ (H Hilbert space). This again is non trivial!

Question: Is the above algebra a Q-algebra?

N. Varopoulos

27.

Let A be a uniform algebra on a compact Hausdorff space X. Assume that for every $M \in A^{\perp}$ (the annihilator of A) the natural injection $i_{\mu} : A \to H_A^1(|\mu|)$ is compact.

Does this imply that $A = C(X)$?

A. Pełczyński

28.

Is it true that every bounded linear operator from the disc algebra $A(D)$ to a Hilbert space can be extended to the space $C(\partial D)$?

A. Pełczyński

29.

Is it true that every operator from $A(D)$ to H^1 can be factorized through a Hilbert space?

A. Pełczyński

30.

Is the space $H^1(D \times D)$ isomorphic to a dual Banach space?

A. Pełczyński

31.

Consider the map $T_{m,n}$ from $\mathbb{C}^m \to \mathbb{C}^n$ defined by

$$w_1 = z_1 + z_2 + \dots z_m$$

$$w_2 = z_1^2 + z_2^2 + \dots z_m^2$$

$$\vdots$$

$$w_n = z_1^n + z_2^n + \dots z_m^n$$

restricted to the torus Π^m $(\Pi = \{z : |z| = 1\}$.

Prove: For every n and every $R > 0$ one has, as soon as $m \geq m_0(n,R)$, that the image of Π^m under $T_{m,n}$ covers the ball of radius R in $\mathbb{C}^n$, centered at 0.

Remark: I have a proof, but it involves unpleasant determinant calculations, and wonder if there is a "clean" one. The proposition is equivalent to: Let $a_0 + a_1 z + \dots a_n z^n$ be any polynomial with $a_0 \neq 0$. Then by suitably adding on higher order terms $a_{n+1}z^{n+1} + a_{n+2}z^{n+2} + \dots$, we can always obtain a polynomial with all roots on $\{|z| = 1\}$.

H. Shapiro

32.

Let H_k be the class of harmonic functions $u(z)$ $(|z| < 1)$ satisfying the inequality

$$-\infty < u(z) \leq k(|z|)\,, \qquad (1)$$

$k(r)$ $(0 < r < 1)$ being a positive function, $k(r) \nearrow \infty$, $(1-r)k(r) \searrow 0$ $(r \uparrow 1)$. Let $\tilde{H}_k \subset H_k$ be the smaller class:

$$|u(z)| \leq k(z) \qquad (1')$$

We are interested in the following property of harmonic functions:

Def.: A class K of harmonic functions $u(z)$ $(|z| < 1)$ is said to have the property $(*)$ if for every $u \in K$ the limit

$$\lim_{r\to 1-0} \int_\alpha^\beta u(re^{i\theta})\,d\theta \quad (0 \leq \alpha < \beta \leq 2\pi) \qquad (*)$$

exists and is bounded in α, β.

Theorem (W.K. Hayman, B. Korenblum): For H_k to have the property $(*)$ it is necessary and sufficient that

$$\int_0^1 \sqrt{\frac{k(r)}{1-r}}\,dr < \infty \qquad (2)$$

Question: Can (2) be relaxed for the class $\tilde{H}$ to have the property $(*)$?

B. Korenblum

33.

Let $\mathrm{Re}\,A = \mathrm{Re}\,A(D)$ be the Banach space of real parts of the disc. algebra with the norm

$$\|f\|_{\mathrm{Re}\,A} = \|f + i\tilde{f}\|_A, \quad \tilde{f}(0) = 0$$

Question: What are the sets E of synthesis for $\mathrm{Re}\,A$. i.e. the closed sets E such that

$$\forall\, f \in \mathrm{Re}\,A,\ f = 0 \text{ on } E,\ \exists f_n = 0 \text{ in a neighbourhood of } E$$
$$\text{s.t. } \|f - f_n\|_{\mathrm{Re}\,A} \underset{n\to\infty}{\to} 0\,.$$

I know it is true if $m(E) = 0$, if E is a interval etc.

J. Detraz

34.

Let $K \subset \mathbb{C}$ be compact with nonempty fine interior K' and let x_0, x be two distinct points in the same fine interior component of K'.

Then $d\nu_x \ll d\nu_{x_0}$, where ν_x, ν_{x_0} are the Keldysh measures for x and x_0. On the other hand, one can find examples where

$$\frac{d\nu_x}{d\nu_{x_0}} \notin L^\infty(\nu_{x_0}).$$

Problem: Give conditions which insure that

$$\frac{d\nu_x}{d\nu_{x_0}} \in L^p(\nu_{x_0}) \quad \text{where } 1 < p \leq \infty.$$

B. Gaveau